ALL THE WORLD'S ANIMALS
CARNIVORES

ALL THE WORLD'S ANIMALS
CARNIVORES

TORSTAR BOOKS
New York · Toronto

CONTRIBUTORS

TNB Theodore N. Bailey PhD US Fish and Wildlife Service Kenai, Alaska USA	**SH** Stephen Harris PhD University of Bristol England	**RAP** Roger A. Powell PhD North Carolina State University Raleigh, North Carolina USA
SKB Simon K. Bearder PhD Oxford Polytechnic England	**HJH** Berty J. van Hensbergen BA Centro Pirenaico de Biologia Experimental, Jaca, Spain/ University of Cambridge England	**PRKR** Philip K. R. Richardson McGregor Museum Kimberley South Africa
BCRB Brian C. R. Bertram PhD Zoological Society of London England	**GK** Gillian Kerby BA University of Oxford England	**JR** Jon Rood PhD Conservation and Research Center National Zoological Park Front Royal, Virginia USA
JDSB John D. S. Birks BSc Universities of Exeter and Durham England	**CMK** Carolyn M. King DPhil Eastbourne New Zealand	**JKR** James K. Russell PhD Formerly of National Zoological Park Smithsonian Institution Washington DC USA
AB Anders Bjärvall PhD National Swedish Environment Protection Board Solna Sweden	**JWL** Jack W. Lentfer Department of Fish and Game Juneau, Alaska USA	
WDB W. D. Bowen PhD Northwest Atlantic Fisheries Center St John's, Newfoundland Canada	**TRL** T. R. Loughlin National Marine Mammal Laboratory Seattle, Washington USA	**MS** Mel Sunquist PhD Conservation and Research Center National Zoological Park Front Royal, Virginia USA
FB Fred Bunnell PhD University of British Columbia Vancouver, British Columbia Canada	**DWM** David W. Macdonald MA DPhil University of Oxford England	**DRV** Dennis R. Voigt MS Ministry of Natural Resources Maple, Ontario Canada
JMD James M. Dietz MS PhD Michigan State University East Lansing, Michigan USA	**AJM** Audrey J. Magoun University of Alaska Fairbanks, Alaska USA	**CW** Chris Wemmer PhD Conservation and Research Center National Zoological Park Front Royal, Virginia USA
ND Nicole Duplaix PhD TRAFFIC Washington DC USA	**JM** James Malcolm PhD University of Redlands Redlands, California USA	
GWF George W. Frame PhD Utah State University Salt Lake City, Utah USA	**PDM** Patricia D. Moehlman PhD University of Yale Newhaven, Connecticut USA	**WCW** W. Chris Wozencraft BSc University of Kansas Lawrence, Kansas USA
FHH Fred H. Harrington PhD Mount Saint Vincent University Halifax, Nova Scotia Canada	**KGVO** Karl G. Van Orsdol PhD University of Cambridge England	**EZ** E. Zimen PhD University of Saarbrücken West Germany

ALL THE WORLD'S ANIMALS CARNIVORES

TORSTAR BOOKS INC.
300 E. 42nd Street,
New York, NY 10017

Project Editor: Graham Bateman
Editors: Peter Forbes, Bill MacKeith, Robert Perberdy
Art Editor: Jerry Burman
Picture Research: Linda Proud, Alison Renney
Production: Bob Christie
Design: Chris Munday

Originally planned and produced by:
Equinox (Oxford) Ltd
Mayfield House, 256 Banbury Road
Oxford, OX2 7DH, England

Editor
Dr David Macdonald
Animal Behaviour Research Group
University of Oxford
England

Artwork Panels
Priscilla Barrett

Library of Congress Cataloging in Publication Data

Main entry under title:

Carnivores.

(All the world's animals)
Bibliography: p.
Includes index.
1. Carnivora. I. Series.
QL737.C2C34   1985   599.74   85-978
ISBN 0-920269-73-7

On the cover: Cheetah female and cubs
page 1: Lion and lioness   *pages 6–7*: Leopard
pages 2–3: Jaguar   *pages 8–9*: Raccoon
pages 4–5: Polar bear

Printed in Belgium

ISBN 0-920269-72-9 (Series: All the World's Animals)
ISBN 0-920269-73-7 (Carnivores)

In conjunction with *All the World's Animals*
Torstar Books offers a 12-inch raised
relief world globe.
For more information write to:
Torstar Books Inc.
300 E. 42nd Street
New York, NY 10017

CONTENTS

FOREWORD

The world of the carnivores embraces creatures in whom resides the very essence of power, endurance and speed, of wit and fang. The lion is the symbol of majesty and authority, the wolf of tirelessness, the fox of guile. Other images of savagery, menace and treachery – whether deserved or not – are no less vivid and are painted here in a rich palette of words and pictures. *Carnivores* is the book that probes the real life of the lion, King of the Beasts, of the tiger and leopard, fox and jackal, bear, raccoon, badger, otter and others of their realm. Not least of these creatures is the giant panda, whose ever-dwindling numbers epitomize man's ruination of the world's wild places.

Carnivores is an ambitious, exciting journey into the kingdom of animals that will appeal alike to the professional and the schoolchild. The newest information and ideas are presented lucidly and entertainingly, but are far from being slight or superficial versions of the truth. The international panel of experts, whose work forms the heart of this book, have seen to it that the stories they tell are not only full of intrigue and incident but meet the highest scientific standards.

Dramatic, dynamic photographs and color drawings bring the text of this volume to life. With their aid, even if you have never seen such creatures in action, it is easy to imagine yourself face to face with a Grizzly bear or the mute spectator of a lion pride at the kill. Here the poetic notions of the animal world are viewed afresh and invested with the crown of knowledge.

How this book is organized

Animal classification, even for the professional zoologist, can be a thorny problem, and one on which there is scant agreement between experts. This volume has taken note of the views of many taxonomists but in general follows the classification of Corbet and Hill (see Bibliography) for the arrangement of families and orders.

This volume is structured at a number of levels. First, there is a general essay highlighting common features and main variations of the biology (particularly the body plan), ecology and behavior of the carnivores and their evolution. Second, essays on each family highlight topics of particular interest, but invariably include a distribution map, summary of species or species groupings, description of skull, dentition and unusual skeletal features of representative species and, in many cases, color artwork that enhances the text by illustrating representative species engaged in characteristic activities.

The main text of *Carnivores*, which describes individual species or groups of species, covers details of physical features, distribution, evolutionary history, diet and feeding behavior, as well as their social dynamics and spatial organization, classification, conservation and their relationships with man.

Preceding the discussion of each species or group is a panel of text that provides basic data about size, life span and the like. A map shows its natural distribution, while a scale drawing compares the size of the species with that of a six-foot man. Where there are silhouettes of two animals, they are the largest and smallest representatives of the group. Where the panel covers a large group of species, the species listed as examples are those referred to in the text. For such large groups, the detailed descriptions of species are provided in a separate Table of Species. Unless otherwise stated, dimensions given are for both males and females. Where there is a difference in size between the sexes, the scale drawings show males.

As you read these pages you will marvel as each story unfolds. But as well as relishing the beauty of these carnivores, you should also be fearful for them. Again and again, authors return to the need to conserve species threatened with extinction and by mismanagement. Of the 231 species described in these pages, one-third are listed in the Appendices I through III of the Convention on International Trade in Endangered Species of Wild Flora and Fauna (CITES). The *Red Data Book* of the International Union for the Conservation of Nature and Natural Resources lists 37 of these species at risk. In *Carnivores*, the following symbols are used to show the status accorded to species by IUCN at the time of going to press. Ⓔ = Endangered—in danger of extinction unless casual factors are modified (these may include habitat destruction and direct exploitation by man). Ⓥ = Vulnerable—likely to become endangered in the near future. Ⓡ = Rare, but neither endangered nor vulnerable at present. Ⓘ = Indeterminate—insufficient information available, but known to be in one of the above categories. ⍰ = Suspected, but not definitely known to fall into one of the above categories. The symbol ⊡ indicates entire species, genera or families, in addition to those listed in the *Red Data Book*, that are listed by CITES. Some species and subspecies that have ⒺⓍ or probably have ⒺⓍ? become extinct in the past 100 years are also indicated.

CARNIVORES

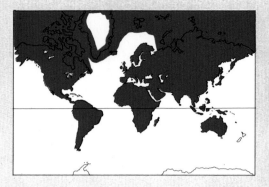

ORDER: CARNIVORA
Seven families; 93 genera; 231 species.

Cat family
Family Felidae—felids
Thirty-five species in 4 genera.
Includes **lion** (*Panthera leo*), **jaguar** *P. onca*),
leopard (*P. pardus*), **tiger** (*P. tigris*), **cheetah**
(*Acinonyx jubatus*), **lynx** (*Felis lynx*), **ocelot**
(*F. pardalis*), **Margay cat** (*F. wiedi*), **Wild cat**
(*F. silvestris*).

Dog family
Family Canidae—canids
Thirty-five species in 10 genera.
Includes **jackals** (*Canis adustus, C. aureus,
C. mesomelas*), **coyote** (*C. latrans*), **wolf**
(*C. lupus*), **Arctic fox** (*Alopex lagopus*), **Red fox**
(*Vulpes vulpes*), **Swift fox** (*V. velox*), **dhole** (*Cuon
alpinus*), **African wild dog** (*Lycaon pictus*).

Bear family
Family Ursidae—ursids
Seven species in 5 genera.
Includes **Polar bear** (*Ursus maritimus*), **Grizzly**
or **Brown bear** (*Ursus arctos*), **Spectacled bear**
(*Tremarctos ornatus*).

Raccoon family
Family Procyonidae—procyonids
Seventeen species in 8 genera.
Includes **raccoon** (*Procyon lotor*), **coati** (*Nasua
narica*), **Giant panda** (*Ailuropoda melanoleuca*),
Red panda (*Ailurus fulgens*).
The Giant panda, and sometimes the Red
panda, are sometimes separated as the family
Ailuropodidae.

Weasel family
Family Mustelidae—mustelids
Sixty-seven species in 26 genera.
Includes **weasel** (*Mustela nivalis*), **Least weasel**
(*M. n. rixosa*), **Black-footed ferret** (*M. nigripes*),

THE Polar bear is up to 25,000 times heavier than the Least weasel; the Giant panda ambles about foraging for bamboo shoots, while the cheetah dashes at up to 60 miles an hour in pursuit of antelope; the aquatic Sea otter and the arboreal Palm civet rarely touch terra firma. Such diversity of form, function and habitat is found throughout the order Carnivora. Represented among the carnivores are the long and thin, the short and fat, the agile and the slow, the powerful and the delicate, the solitary and the sociable, the predatory and the preyed-upon. There is little in the outward appearance of the 231 species, 93 genera and seven families to unite them.

So what does distinguish the Carnivora from other mammals? Ultimately their common lineage rests on one shared characteristic—the possession of the four so-called carnassial teeth. Many species in other orders, past and present, have been meat-eaters, but only members of the Carnivora stem from ancestors whose fourth upper premolar and first lower molar were adapted to shear through flesh. Only the more predacious of the modern species retain this pair of slicing teeth, collectively called carnassials. In species with more vegetarian inclinations, such as pandas, they have reverted to grinding surfaces.

The Ferocious Image
Many living Carnivora are adapted to either mixed (omnivorous) or even largely vegetarian diets, but meat-eating has been their speciality in the past. Although easier to digest than plant material, animal prey is harder to catch, and much of the fascination of carnivores lies in the stealth, efficiency, precision and the almost unfathomable complexity of their predatory behavior.

Prey are killed in various ways. Civets and mongooses (family Viverridae) and weasels and polecats (Mustelidae) are generally "occipital crunchers." They bite into the back of the head and so smash the back of their prey's braincase. Handling prey is dangerous to predators, and in the cases of these two families the victim's armaments are kept well out of harm's way by highly stereotyped behavior. A weasel, for example, will throw itself on its side or back while delivering the killing bite, pushing away the claws and teeth of the struggling prey with thrusts of its legs. When dealing with small prey, members of the cat family (Felidae—the felids) aim a neck bite which prises apart cervical vertebrae with their sharp-pointed canines. Members of the dog family (Canidae—the canids) generally grab for the nape of the neck, as they tackle small prey, or pinion them to the ground with their forepaws. The grab is followed by a violent, dislocating head shake. However, canids immobilize larger prey through shock, which results from a combination of throat and nose holds, and bites to exposed soft parts that often disembowel the victim.

Carnivores have been dubbed vicious and cruel by people who equate anger and murder in man with social aggression and killing of prey in carnivores; but there is nothing aggressive, let alone vindictive, in a carnivore killing its prey, any more than there is in a herbivore decapitating a plant stem. The lion throttling a zebra is involved in the same function—feeding—as the zebra is in cropping grass.

One predatory phenomenon, more than any other, has resulted in carnivores being unfairly reviled—"surplus killing," that is,

► ▼ **Diet, hunting tactics and habitat** of carnivores vary enormously. In the open grasslands of East Africa, African wild dogs hunt in packs for large prey such as wildebeest OPPOSITE ABOVE which they run down, then kill by a combination of nose and tail holds coupled with disemboweling. The bobcat of North America RIGHT inhabits rough terrain, and hunts alone and by stealth for small prey such as rabbits and mice, which it swiftly dispatches by a bite to the neck or throat. The Giant panda BELOW lives in mountainous forests of China, where its placid search for bamboo "prey" lies at one extreme of carnivoran foraging.

ermine or **stoat** (*M. erminea*), **Eurasian badger** (*Meles meles*), **Striped skunk** (*Mephitis mephitis*), **Marine otter** (*Lutra felina*), **Sea otter** (*Enhydra lutris*).

Civet family

Family Viverridae—viverrids
Sixty-six species in 37 genera.
Includes **Palm civet** (*Paradoxurus hermaphroditus*), **genets** (*Genetta* species), **meerkat** (*Suricata suricatta*), **Indian mongoose** (*Herpestes auropunctatus*), **Dwarf mongoose** (*Helogale parvula*), **Banded mongoose** (*Mungos mungo*).
The mongooses are sometimes separated as the family Herpestidae.

Hyena family

Family Hyaenidae—hyenids
Four species in 3 genera.
Includes **Spotted hyena** (*Crocuta crocuta*), **Brown hyena** (*H. brunnea*).

killing more than they can eat in one meal. The farmer who discovers in his coop the slain bodies of a dozen or more chickens, some of them seemingly needlessly decapitated, will vow the fox kills for pleasure.

Many species of carnivores do indeed engage in surplus killing given the opportunity—wolves in a sheep fold, lion among cattle, Spotted hyenas in a herd of gazelle—all do so. But an alternative explanation to "blood lust" is more consistent with the facts. Prey are elusive, almost as well adapted to avoiding predators as their predators are to catching them; so tomorrow's dinner is never assured for a carnivore. However, since many carnivores cache or store portions of their prey or defend an unfinished carcass, natural selection has favored behavior that enables the predator to make the most of any windfall opportunity to kill unwary prey.

In practice, the prey's avoidance of the predator effectively limits the slaughter. Where this does not happen, and so-called surplus killing occurs, it is almost always because human intervention has compromised escape behavior—shut up in their coop, the chickens flutter frantically, but to no avail, as the fox seizes the opportunity to make extra kills.

The Body Plan of a Carnivore

Amongst the carnivores are representatives of almost every variation on the mammalian theme. However, the skeletons of all carnivores, irrespective of whether they walk on the soles of their feet (plantigrade), for example, as bears do, or on their toes (digitigrade) as do canids, share an evolutionarily ancient modification of the limbs—the fusion of bones in the foot (see

BODY PLAN OF A CARNIVORE

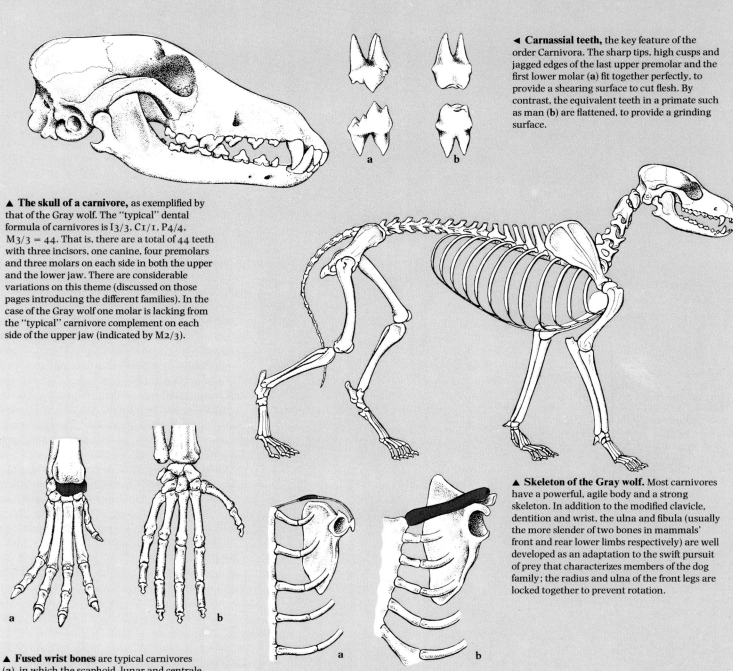

◄ **Carnassial teeth,** the key feature of the order Carnivora. The sharp tips, high cusps and jagged edges of the last upper premolar and the first lower molar (**a**) fit together perfectly, to provide a shearing surface to cut flesh. By contrast, the equivalent teeth in a primate such as man (**b**) are flattened, to provide a grinding surface.

▲ **The skull of a carnivore,** as exemplified by that of the Gray wolf. The "typical" dental formula of carnivores is I3/3, C1/1, P4/4, M3/3 = 44. That is, there are a total of 44 teeth with three incisors, one canine, four premolars and three molars on each side in both the upper and the lower jaw. There are considerable variations on this theme (discussed on those pages introducing the different families). In the case of the Gray wolf one molar is lacking from the "typical" carnivore complement on each side of the upper jaw (indicated by M2/3).

▲ **Skeleton of the Gray wolf.** Most carnivores have a powerful, agile body and a strong skeleton. In addition to the modified clavicle, dentition and wrist, the ulna and fibula (usually the more slender of two bones in mammals' front and rear lower limbs respectively) are well developed as an adaptation to the swift pursuit of prey that characterizes members of the dog family; the radius and ulna of the front legs are locked together to prevent rotation.

▲ **Fused wrist bones** are typical carnivores (**a**), in which the scaphoid, lunar and centrale bones are fused together to form the scapholunar bone; in a primate (**b**), these bones remain independent.

▲ **The collar bone is reduced** in all carnivores ABOVE RIGHT (**a**) by comparison with other mammals, such as a primate (**b**). Shown here is the collar bone (clavicle) of a wolf, which is reduced to a mere sliver of bone (*red*) suspended on ligaments (*blue*), and of a man.

► **Jaw power** is crucial for the capture and tearing up of prey. Shown here are the lines of force exerted by the jaw-closing muscles of the dog. The massive temporalis (**a**) delivers the power to exert suffocating or bone-splitting pressure, even when the jaws are agape; the rearmost (posterior) fibers of the muscle are most effective when the jaws are open wide. The masseter muscle (**b**) provides the force needed to cut flesh, and for grinding when the jaws are almost closed.

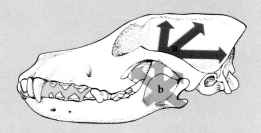

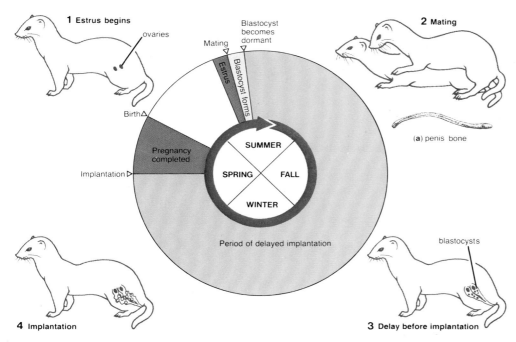

1 Estrus begins

ovaries

Mating

Blastocyst becomes dormant

Estrus

Blastocyst forms

Birth △

Pregnancy completed

Implantation ▷

SUMMER

SPRING · FALL

WINTER

Period of delayed implantation

2 Mating

(a) penis bone

blastocysts

4 Implantation

3 Delay before implantation

▲ **Reproductive cycle of the stoat** or ermine (family Mustelidae), one of a number of variations on the normal mammalian cycle displayed by some carnivore species.
(1) **Estrus begins.** In early summer, females come into estrus (heat) and up to 10 eggs mature in the ovaries.
(2) **Mating** soon occurs and the eggs are released as a result of copulation (*induced ovulation*), through hormonal and nervous stimulation. The prolonged copulation necessary to trigger ovulation is facilitated by the *penis bone* or *baculum* (**a**, shown actual size) which supports the penis of the male.
(3) **Delay before implantation.** After fertilization the single-celled eggs travel to the uterus, each developing as it goes into a ball of cells – the blastocyst. Instead of implanting in the uterus wall and continuing to grow into the embryo (as normally happens in most mammals), the blastocysts continue to float free and do not develop further. This *delayed implantation* (or embryonic diapause) lasts for a few days in some species, up to 10 months in the case of the stoat. The delay is maintained by low levels of the hormone progesterone.
(4) **Implantation.** In the spring, increased day length, as perceived by the eye, affects the hormone balance, and progesterone levels increase. This allows the blastocysts to implant; pregnancy is completed in 28 days.

OPPOSITE). For members of the dog family, the development of this "scapho-lunar" bone might plausibly be interpreted as providing a firm strut for absorbing the shock of landing at the end of a limb adapted for running. However, the fused bones were present in older, now extinct, forest-dwelling carnivores, so perhaps the scapho-lunar bone originally provided a firm basis for flexion at the mid-carpel joint in carnivores which needed both to climb well and to grapple with prey.

The advantages of a long stride when running may explain the relatively undeveloped collar bone (clavicle), free at both ends and lodged within shoulder muscles (see LEFT). The main function of a large clavicle is to stabilize the lower end of the shoulder blade and to provide attachment for the muscles controlling side-to-side movement of the limbs; neither function is either necessary or desirable for the fore-and-aft swing of the limbs of carnivores, which primarily run down prey.

Once caught, prey must be killed and dismembered. These two quite distinct tasks are served by two sets of jaw muscles, each of which exerts most force at a different phase of the bite. The temporalis muscle runs from the top (coronoid) process of the lower jaw to the side of the braincase behind the eye. This muscle is most effective when the jaws are wide open and is vital for the application of the killing stab of the canine teeth. For those species, such as the big cats, with large canine teeth, more skull surface is required as a base for relatively huge temporalis muscles: this is the function of the so-called sagittal and parietal crests that run

along the top of the skulls of these species. The masseter muscles are used to grind and cut food, when the jaws are virtually closed. The masseter runs from the lower (angular) process of the jaw to the zygomatic arch below the eye. The relative development of these sets of muscles (and the shape of the skull) depends on the species' life-style.

The remainder of carnivore anatomy is just as varied as their diverse life-styles would lead us to expect. The retractile claws so typical of cats are not common to all carnivores; apart from cats they are found only amongst the Viverridae (some civets and genets). Canids, by contrast, have digging claws which they use in caching food. The forefeet and hindfeet generally have four and five digits respectively.

A few unusual reproductive traits are found in carnivores. The penis of members of all families, except the hyenas, contains an elongate bony structure known as the baculum or *os penis*. This penis bone functions to prolong copulation, which may be especially important in species where ovulation is induced by copulation. The shape of the baculum is characteristic for each species. The so-called copulatory tie, which "locks" male and female together during copulation, occurs only in canids (see p48).

In most species the cycle of development of the fertilized egg is typical of that found in most mammals—the egg develops continuously from fertilization to birth of the offspring. In some carnivores, principally members of the weasel family, a delay in development occurs (see previous page; also p111).

The senses of the carnivores are all acute. Perhaps most intriguing is their refined ability to use scent not only to find prey (and to escape predators) but as a method of communication. Apart from the signal value of urine and feces which are deployed at strategic locations, most carnivores have several odorous skin glands. Doubtless these odors convey far more complex information than we can yet confirm. It is already known, for example, that one mongoose can recognize the identity and status of another by scent alone and it is likely that most carnivores can recognize individually others of their species from their scent marks.

Distribution

Wild carnivores have a worldwide distribution, with the Arctic foxes of Greenland and the feral cats of Subantarctic Marion Island at the extremes of their latitudinal range. Tundra wolves, rain forest civets, marine otters and desert foxes are among those species which illustrate how carnivores have adapted to every major habitat. Each family is widespread, the dog family most of all. Some species have been introduced by man to areas where they are not native, generally with disastrous consequences; Small Indian mongooses shipped to the Caribbean to control rats spread rabies instead; feral cats imported to remote islands annihilate flightless birds; whereas stoats and Red foxes introduced to control rabbits and to provide sport, in New Zealand and Australia respectively, actually threaten the native faunas.

There is one recent and intriguing move that the carnivores have undertaken voluntarily, and that is into the urban environment. The rubbish tips of the Middle East have provided welcome tidbits for jackals since biblical times, skunks and raccoons forage in the suburbs of North American cities, and the Spotted hyenas that wander the streets of Harer in Ethiopia are widely reported. In the last 20 years carnivores have been knocking at the gates to the capitals of Europe; Eurasian badgers now occupy setts in London and Copenhagen, and Red foxes are seen by lamplight in the streets of Stockholm, Copenhagen, Paris, London and many other towns besides. For hitherto unexplained reasons, urban foxes and badgers are most established in the United Kingdom and, in the case of foxes at least, in the southeast and the Midlands of England especially.

In the built-up districts of Bristol, England, the majority of badger setts are dug in private gardens or isolated strips of woodland. The badgers forage within ranges of 77–200 acres (31–81 hectares) for the varied diet, including earthworms, typical of some rural badgers. However, while rural badgers have so far proven to be rather strictly territorial, the ranges of urban badgers overlap widely (see Societies and Social Behavior, right). Red foxes are common around the university buildings and throughout the city of Oxford. There, the vixen who reared cubs in an automobile factory competes in notoriety with foxes seen in London's Trafalgar Square and Waterloo Rail Station! Oxford's foxes travel ranges that average 213 acres (86 hectares) and encompass every urban habitat from terraced housing to ornamental gardens. They may supplement the rural fox's diet with scraps from bird tables and compost heaps, and with the smaller casualties of road traffic. Only rarely do they raid and overturn trash cans (the culprits are usually dogs).

The Evolution of Carnivores

The early mammals are known largely from their teeth, since the smallness and fragility of their bodies, and also their forest habitats, have not favored the preservation of complete fossil remains. Consequently, we have only fragmentary knowledge of the origins of mammals 190 million years ago in the Tertiary era, and also of the ancestors of modern mammalian orders about 70 million years ago. Among the ancient carnivorous types a specialized pair of shearing carnassials evolved independently several times, for example, in the now extinct order of Creodonts, in which they evolved in different parts of the tooth row from those of modern carnivores.

The most likely forerunners of all living Carnivora are members of the extinct superfamily Miacoidea. These poorly known forest dwellers had spreading paws, probably indicating a tree-dwelling life-style, and carnassials derived from the fourth upper premolar and first lower molar, although the scaphoid and lunar bones were not yet fused. From the miacids the modern carnivore families developed during a fast radiation in the Eocene and Oligocene periods (54–26 million years ago). Doubtless this proliferation of types of predator mirrored a similar evolutionary explosion of potential prey, which in turn developed from the availability of more diverse vegetable food.

Among the more dramatic histories of modern carnivore families is that of the Felidae. All present-day cats are classified within the subfamily Felinae, but their early days were overshadowed by the successful radiation of saber-toothed cats of the subfamily Machairodontinae, which dominated the felid scene from the Miocene to the Pliocene periods (26–2 million years ago). Whereas the lower canines of modern cats are only slightly smaller than their upper ones, the lower canines of saber-toothed cats were reduced to vestiges in order to make room for their massive counterparts in the upper jaw. The extinct genus *Hometherium* had blade-like upper canines, serrated along their inner edge, whereas those of the American Pleistocene genus *Smilodon* were much longer and conical, fashioned more for piercing than for cutting. Probably cats of the *Homotherium* type severed major blood vessels, trachea and esophagi with a rending throat bite to their prey, whereas *Smilodon* was specially adapted to stabbing the throat of thick-skinned victims. It seems unlikely that either type of saber-toothed cat used its long, and hence fragile, fangs to prize apart the vertebrae of

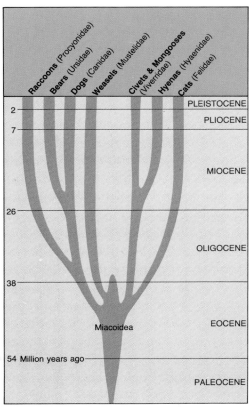

▲ **Evolution of the carnivore families.** Rapid differentiation of mammal, insect and flowering plant species has characterized most of the Cenozoic ("new animal") era of the earth's history. Most of the seven families of the order Carnivora became distinct in the first half of the era, although two appeared later, in the Miocene (bears and hyenas). Saber-toothed cats (see text) were just one example of a tendency to gigantism in the Pliocene and Pleistocene periods. Many large carnivores have died out in relatively recent times, and today smaller species predominate in most families.

▼ **The urban carnivore.** While man disturbs or even destroys their natural habitat many carnivores penetrate our cities in search of food. Prominent among such species in northern Europe is the Red fox, caught here while on a night-time foray in the suburbs of Oxford, England.

► ▼ **Social interactions in the Red fox,** one of several species of carnivore previously supposed to be solitary and which turns out to have a complex social life. The dominant vixen of a Red fox group defers RIGHT to the dominant male (standing) by lying prostrate in his path, yet still manages to intimidate a subordinate female, her sister, who crouches and gapes submissively with ears folded back. BELOW A male solicits play from a vixen by rolling on his back and tugging playfully at her throat. BOTTOM Two adult sisters who have always shared the same home range sit amicably together, one grooming the other.

prey as do the small and medium-sized cats of today. More probably they would bite the victim's throat, as do modern big cats when tackling large prey. The exaggerated size of these teeth presented a severe mechanical problem. If the mouth was to open sufficiently wide to deliver the *coup de grâce*, the action would have involved risk of the predator dislocating its own neck. Consequently, the front neck muscles of Machairodontinae were enormous, and inserted farther beneath the skull than in today's cats, to give optimum mechanical advantage during the stab.

Societies and Social Behavior

Refined though the anatomical specializations of the carnivores may be, and however elegant the details of their behavior, the overwhelming feature of their biology is the subtlety of their societies. The collective strength, coordinated strategy and awesome effectiveness with which the cooperative hunters overpower their prey has captured the human imagination. But there is much more than cooperative hunting, spectacular though it is, to the societies of carnivores who hunt together, and it is increasingly clear that some carnivores have quite different yet equally complex societies whose origins and maintenance have nothing to do with cooperative hunting.

Traditionally, two ideas were advanced to explain why some carnivores go around in groups: some, such as wolves and lions, it was argued, hunt together only in order to cooperate in the capture of large, dangerous quarry; others, such as some mongooses, were thought to travel together just to enjoy greater vigilance for marauding predators (see pp139, 144), some of which they could collectively repel. Certainly both these are among the chief selective pressures which have fashioned some carnivore societies, but neither is an appropriate explanation of why other species live in groups but travel and hunt alone. There are several species whose shy, nocturnal habits and small prey previously misled people into thinking them strictly solitary or asocial, for example foxes,

civets, Brown hyenas, farm cats and Eurasian badgers. The use of radio-tracking and night vision equipment have now revealed that in each of these species (and probably many more) several adults may share roughly the same home range, even if the cohabitants meet only infrequently when foraging, and sometimes even den separately. Such species may be said to live in "spatial groups" whose members' home ranges overlap more than would be expected by chance. Three species which show how other pressures may favor the formation of spatial groups, in the *absence* of concerted hunting or anti-predator behavior are the Eurasian badger and Red and Arctic foxes.

In much of rural Britain the Eurasian badger lives in group territories within which 2–10 animals den together but forage alone, principally hunting for the earthworm *Lumbricus terrestris*. At night this species of worm only crawls from its burrow when the grass temperature is over 36°F (2°C), the air calm and humidity high. The problem for the badgers is that the worms emerge in different places from one night to the next, depending on slight variations in weather conditions. So, to be sure of finding worms, a badger requires access to a territory large enough to accommodate such a variation of climate and therefore worm availability. However, in the night's "good patch" many more worms are often available than one badger can possibly consume, so at no personal cost it can tolerate the presence of others in its territory. From this idea the scientist Hans Kruuk developed a model which correctly predicted that, where patches of earthworms were scattered, badger territories would be bigger, and, independently, that where patches were richer in worms, social groups of badgers would be larger.

The life-styles of Red and Arctic foxes share this feature of foraging for "patchily" dispersed prey, and the "badger model" may apply: territory size is determined by the dispersion of good feeding sites but group size by the amount of food at each feeding patch. For them, as with the badger, there may be little cost to tolerating some additional group members so long as food is plentiful, and, furthermore, in each species positive advantages of group formation have been discovered. Red foxes on the outskirts of Oxford forage in lush gardens where fruit, small mammals, invertebrates and household scraps are abundant. The gardens represent rich patches amidst relatively barren farmland. These suburban foxes live in groups of 3–6 adults which occupy territories varying from 47–178 acres (19–72 hectares). Each territory contains a similar number of gardens (about 24) and includes enclaves of housing; where these enclaves are widely dispersed the territories are larger. Similarly, in the fjords of Iceland, Arctic foxes in groups averaging three adults occupy coastal territories of 3.3–7.1 sq mi (8.6–18.5 sq km). Sixty to 80 percent of their food is obtained by beachcombing, and only beaches favored by the drift of sea currents are bountiful. Each cove thus constitutes a potential foraging patch, into which carrion and driftwood may be swept by the tide. The farmers in this part of the world collect all the driftwood to make fence posts. In each of the fox territories measured, irrespective of the length of coastline it encompassed, the farmers found wood to make roughly 1,900 fence-posts; presumably, therefore, equal supplies of carrion for the foxes were also washed ashore. In both Red and Arctic fox societies a single dog fox forms a spatial group with several related adult vixens of which generally only one, probably the mother of the rest, gives birth to cubs. As the cubs mature a conspicuous advantage of tolerating non-breeding vixens becomes evident: some non-breeders help to feed the cubs, and spend time grooming and playing with them and, presumably, keeping a watchful eye open for danger. This phenomenon of "helping" (alloparenthood) is widespread amongst carnivores (see RIGHT), especially among members of the dog family.

Other carnivores, such as lions, African wild dogs, dholes, wolves, jackals and coyotes also cooperate in the care of young, as well as benefiting from cooperative hunting. Similarly, members of the four genera of mongooses which benefit from group defense against predators also show allo-parental behavior. In all cases individuals contribute to, and benefit from, various aspects of group life to different extents: in a pride of lions the males join forces to repel rivals, but hunt less than the lionesses who also suckle each other's cubs (see pp26–27). And within a pack of dholes some individuals may regularly lead the hunt, while others are especially vigilant guards at the den (see pp72–73). The pros and cons of group membership vary also between individuals as their role in the group alters, depending on sex, age and status. In the same way the suite of advantages that makes group living advantageous varies from species to species, and the nature of their societies varies accordingly. The lesson

is that no one selective pressure is the sole force for any carnivore species' sociality.

In addition to cooperative hunting, vigilance and infant care, other advantages of group living are becoming apparent: larger groups of coyotes (see pp54–55) and Golden jackals can better defend their prey from rival groups, as can bigger parties of Spotted hyena defend theirs from marauding lions; Dwarf mongooses collaborate in the care of an ailing group member: in the lion and cheetah (see pp34–35) coalitions of males roam together with better chances of usurping resident males than they would have alone; Gray meerkats take it in turns to climb to a vantage point while their companions feed safely; and hunting tricks pass from one generation to another in packs of African wild dogs (see p69) and lion prides.

With variations on these themes, the list is still growing. In all cases, however, ecological circumstances, such as the pattern of food availability, set the limits for what is feasible socially.

Man and Carnivores

Man's relationship with carnivores is one of extremes—the Domestic dog and cat are to

Shared Parenthood

This non-breeding Gray meerkat male is caring for his three-week-old siblings while their mother is away foraging. Both male and female helpers guard the young at the den, a substantial sacrifice since they may not forage all day. Until the babies are weaned their mother neither guards nor provisions them with prey, but puts all her energies into foraging for herself, so as to produce abundant milk. Later, helpers also provision the youngsters with food.

By contrast, in a pack of Banded mongooses, maternal duties are shared among several breeding females, as they are in a band of coatis, and by both females if two Brown hyenas breed together. Lions and domestic cats, but not other felids, commonly live in close-knit groups of related females who share the nursing of each other's cubs.

Within the dog family, although both do happen, joint denning of litters and nursing coalitions are the exception rather than the rule; in most canids reproduction is the prerogative of the most dominant pair within each group. Subordinate animals defer reproduction, perhaps indefinitely, but more often until they accede to dominance within their natal group or disperse to establish a new group. The subordinates' postponement of reproduction results from social suppression, sometimes through direct violence, by the dominant pair. The process can apply to canids which travel and forage as

a pack, such as wolves, dholes and African wild dogs, or to those who often travel and forage alone, such as jackals and foxes. In each of these species at least some non-breeders may help tend the breeder's offspring, provisioning them with food. In Red fox groups, whose subadult males emigrate, the helpers are invariably female, while in jackal groups, and probably in those of other members of the genus *Canis* (eg wolves and coyote), males and females are equally likely to be helpers. In the "back-to-front" society of African wild dogs it is young females who emigrate and the majority of helpers are male (see pp70–71).

There are several possible reasons why non-breeding animals may tend the offspring of others: they may be acquiring practice at parenthood, or may benefit subsequently from increased group size or (since groups are often composed of kin) may be investing in infants with which they share almost as many genes as they would with their own offspring (see p58).

▲ ▼ Oil fields are the last refuge of the San Joaquin kit fox in Central Valley, California, of which under 7,000 survive. The remainder of its former range has been destroyed by cotton farms. Where oilmen collaborate with conservation agencies to route new roads away from breeding dens, this fox suffers no ill-effects from the increase in industrial activity. Despite the presence of 100–220 wells per sq mile (40–80 per square km), breeding adults of this endangered subspecies of Swift fox survive at densities of 2–2.8 per sq mile (0.8–1.1 per square km). The survival of other carnivores depends on such acceptance of the shared costs and benefits of integrated land use.

The small black patch visible at the root of the tail BELOW is the site of the violet (or supracaudal) gland—a large scent gland of mysterious function present in most canids.

be found in their millions in all corners of the globe, while some wild species have had their numbers reduced to hundreds and others have been completely annihilated. The dog was one of the first animals to be domesticated, the origins of its close relationship with humans going back some 14,000 years, when man mainly lived in hunter-gatherer societies. It is now generally agreed that the wolf is the ancestor of the Domestic dog, but it is still debated whether dogs were deliberately domesticated to serve as hunters, guards or scavengers, or as sources of food or for warmth at night, or were adopted as pets or companions. The ancestry of the domesticated cats dates back no more than 4,000 years (possibly only 3,500 years) and there is no indication that they were domesticated for any practical purposes.

In common with many groups of mammals, numerous members of the Carnivora are threatened by man either directly through persecution and exploitation or indirectly through destruction of their habitat. Even if in no immediate danger of extinction, almost all carnivores require conservation in the sense of thoughtful management, since their maligned reputations, as much as their predatory behavior, have turned rural people against them. Despite a generally open verdict on

whether predator control is beneficial for either stock protection or disease (rabies) control, more than one Red fox is killed annually per square kilometer (two per square mile) over much of Europe. The onslaught by stockmen on the coyote of North America is notorious (see p55), and in the USSR, following an estimated annual loss of one million cattle to wolves in the 1920s, a precedent was set for killing up to 40,000–50,000 wolves annually. Today wolves are still bountied in the USSR (female plus pups for 200 rubles), with the result that 32,000 were killed in 1979. It is ironic that meanwhile biologists elsewhere struggle to secure the survival of the tiny relict populations of wolves in Italy (about 100 animals), Poland (about 200), Portugal (about 100), Egypt (about 30) and Norway (less than 10 animals).

As their habitats dwindle and populations become more fragile the fate of many species of carnivore is totally in human hands. Our society must decide whether these fascinating and often strikingly beautiful creatures are to survive or not. The problem is illustrated by the fact that the same Red fox, for example, may be seen by different people as aesthetically stunning, as a rabies vector, a noble (if inedible) quarry, a killer of lambs or pheasants, a "useful" predator upon rodents, or a pelt to be harvested.   DWM

THE CAT FAMILY

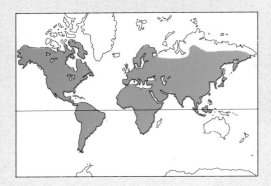

Family: Felidae [*]
Thirty-five species in 4 genera.
Distribution: every continent and major island except Australia, Madagascar and Antarctica. Domestic cat on all continents.

Habitat: very diverse, ranging from desert, through forest to mountain areas.

Size: head-body length from 14–16in (35–40cm) in the Black footed cat to 7.8–10.2ft (2.4–3.1m) in the tiger; weight from 2.2–4.4lb (1–2kg) in the Black-footed cat to 845lb (384kg) in the tiger.

[*] CITES listed

Big cats
Seven species.
Lion *Panthera leo*
Tiger *P. tigris*
Leopard *P. pardus*
Jaguar *P. onca*
Snow leopard *P. uncia*
Clouded leopard *Neofelis nebulosa*
Cheetah *Acinonyx jubatus*

Small cats
Twenty-eight species of *Felis*, including **lynx** (*F. lynx*), **bobcat** (*F. rufus*), **puma** (*F. concolor*), **Wild cat** (*F. sylvestris*, includes **Domestic cat**), ocelot (*F. pardalis*), and **Black-footed cat** (*F. nigripes*).

Felids (family Felidae) are the most carnivorous of the order Carnivora, feeding almost exclusively on vertebrate prey; they sit at the pinnacle of many food pyramids and most have few predators apart from man.

Early forms of modern cats (subfamily Felinae) can be dated to the Miocene (25 million years ago), although the earliest felids evolved in the Eocene (50 million years ago), while the saber-toothed cats (later genera *Homotherium* and *Smilodon*) originated in the intervening Oligocene (38 to 26 million years ago) (see p14). A fundamental distinction between the big cats of the genus *Panthera* and the small cats (genus *Felis*) is that the big cats can roar but cannot purr whereas the small cats can purr continuously but cannot roar. The big cats' ability to roar stems from the replacement of the hyoid bone at the base of the tongue with pliable cartilage allowing greater freedom of movement.

Density of hair, coloring and patterning of the coat vary considerably in relation to habitat. The basic color is usually a shade of brown, gray, tawny or golden-yellow, and this is often patterned with darker circles, stripes, rosettes or spots. Many species have a dark stripe—the tear stripe—running alongside the nose, from the corner of each eye.

Felids have large eyes with binocular and color vision. In daylight, they see about as well as man, but under poor illumination their sight is up to six times more acute than ours. Their eyes adapt quickly to sudden darkness by rapid action of the iris muscles that control pupil diameter. The image is further intensified by a reflecting layer, the *tapetum lucidum*, which lies outside the receptor layer of the retina. Any light that passes through the receptor layer without being absorbed is reflected back again and may stimulate the receptors a second time.

The outward sign of this is the "eyeshine" seen when light is shone into cats' eyes at night.

Felids have large ears which funnel sound waves efficiently to the inner ear. The small cats are particularly sensitive to high-frequency sound. The felid sense of smell is less well developed than in canids, although it is still important.

As well as smell and taste, there is a third

▼ **The big cats – some characteristic activities.** (1) Snow leopard (*Panthera uncia*) scent marking a rock by projecting urine backward. (2) Clouded leopard (*Neofelis nebulosa*) sharpening its claws on a tree. (3) Lions (*Panthera leo*)—lionesses head-butt in greeting. (4) Cheetahs (*Acinonyx jubatus*) – two subadults play-chasing. (5) Tiger (*Panthera tigris*) roaring, a call that warns other tigers it is in residence. (6) Jaguar (*Panthera onca*) covering prey with leaf litter. (7) Leopard (*Panthera pardus*) caching prey up a tree.

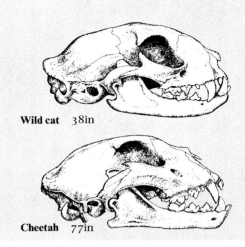

Wild cat 38in

Cheetah 77in

Skulls of Felids

Felid skulls are small, with a short face resulting from reduction in the nasal cavity and jaw length. The dental formula (see p12) is I3/3, C1/1, P3/2, M1/1 = 30, except for the lynx and Pallas's cat, which lack the first

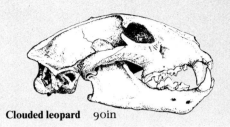

Clouded leopard 90in

upper premolars and hence have 28 teeth. The molars and premolars are adapted as gripping and tearing carnassials. The upper carnassial tooth (premolar 3) has a dual purpose: it has a sharp cutting edge but its anterior cusp is relatively broad and is used to crush bones. The canines are large (particularly so in the Clouded leopard) and used for grabbing and killing prey. Jaw mobility is restricted to vertical movements with the powerful masseter muscle giving a vice-like grip. To compensate for the lack of chewing molars, the tongue is coated with sharp-pointed papillae which retain and lacerate food and rasp flesh off a carcass. Each genus has its own arrangement of papillae.

sense, which utilizes the vomeronasal or Jacobson's organ sited in the roof of the mouth. The use of Jacobson's organ is associated with a distinctive facial gesture, an open-mouthed lip curl known as the Flehmen reaction. This is most commonly seen in a sexual context, possibly helping the two sexes to find one another.

The facial whiskers are long, stiff and highly sensitive; they are especially useful when hunting at night. Like a fingerprint, each cat's whiskers form a unique pattern.

Over three-quarters of cat species are forest dwellers and agile climbers, with a renowned reflex ability to fall on their feet: the vestibular apparatus of the inner ear, which monitors balance and orientation, acts in conjunction with vision to provide information relayed to the brain on the falling cat's orientation. The neck muscles

rotate the head to an upright horizontal position, with which the rest of the body rapidly aligns so that the cat lands upright.

The majority of felids are solitary, extremely secretive and live in inaccessible, remote areas. The only real threat to their survival comes from man, notably the trade in spotted skins, which has brought species such as the tiger, leopard and ocelot near to extinction   GK

LION

Panthera leo [*]
One of 5 species of the genus *Panthera*.
Family: Felidae.
Distribution: S Sahara to S Africa, excluding
Congo rain forest belt; NW India (a remnant
population only in Gir Forest Sanctuary).

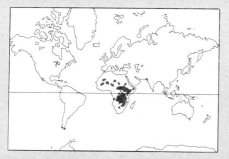

Habitat: varied, from rich grasslands of E Africa
to sands of Kalahari Desert.

Size: male head-body length
8.5–10.8ft (2.6–3.3m); tail
length 2–3.3ft (60–100cm);
shoulder height 4ft (1.2m);
weight 330–530lb
(150–240kg). Female head-
body length 8–9ft
(2.4–2.7m); tail length
2–3.3ft (60–100cm);
shoulder height 3.6ft
(1.1m); weight 270–400lb
(122–182kg).

Coat: light tawny; white on abdomen and
inner side of legs; back of ears black; mane of
male tawny through reddish-brown to black.
Coat of immature animals has a rosette pattern
which fades as they mature, although vestiges
may remain on lower abdomen and legs of
adults.

Gestation: 100–119 days.

Longevity: about 15 years (to 24 in captivity).

Subspecies: 7. **Angolan lion** [E] (*P. l. bleyenberghi*),
Zimbabwe, Angola, Katanga (Zaire). **Asiatic lion** [E]
(*P. l. persica*), Gir Forest, NW India; coat thicker than
African lions with a longer tail tassel, a more
pronounced belly fringe and a more prominent tuft
of hair on its elbows, mane smaller. **Masai lion**
(*P. l. massaieus*), E Africa. **Senegalese lion** [E]
(*P. l. senegalensis*), W Africa. **Transvaal lion** [E]
(*P. l. krugeri*), Transvaal. **Barbary lion** [EX] (*P. l. leo*),
N Africa. **Cape lion** [EX] (*P. l. malanochiata*), Cape to
Natal.

[*] CITES listed.　　[E] Endangered.　　[EX] Extinct.

▷ **Two lionesses prepare to hunt together.**
It is the pride's females that do most hunting.

▶ **A battle-scarred lion** rests in the midday
heat. The main role of males is to protect the
pride from other marauding males. This one has
clearly experienced several such encounters.

B ECAUSE of its strength and predatory
habits, the lion has been considered for
many centuries to be the "King of Beasts."
The myth of the supernatural powers of the
lion survives today: by consuming or wear-
ing parts of a lion it is believed that one can
revive lost powers, cure illness and win
immunity from death. The powerful image
of the creature still lures hunters to Africa
from all parts of the world to demonstrate
their prowess and courage by shooting one;
a trophy bestows social prestige on its
owner. Fortunately, most people are now
content merely to watch or to photograph
this magnificent animal.

Lions were once far more widely distributed than they are today. Cave paintings
and archaeological finds testify to their
widespread presence in Europe some
15,000 years ago. The writings of Aristotle

mention lions in Greece as recently as 300 BC, and the Crusaders frequently encountered lions on their journeys through the Middle East. Lions could still be found in much of the Middle East and northern India up to the turn of the century.

Like other members of the cat family, the lion has a lithe, compact, muscular and deep-chested body. Its head is rounded and shortened and bears prominent whiskers. The skull is highly adapted to killing and eating prey, and the jaws are short and powerful. Backward-curved horny papillae cover the upper surface of the tongue; these are useful both in holding onto meat and in removing parasites during grooming. Vision and hearing are of greater importance than sense of smell in locating prey. As in most other cats, adult male lions are considerably larger than adult females (20–35, sometimes 50, percent heavier). The males' greater size gives them a marked advantage at feeding sites, where they are able to crowd in or even to steal carcasses for themselves. Indeed, pride males may survive almost exclusively on kills made by females.

The male's chief role in the pride is to defend the territory and the females from other males and size is obviously an advantage here too. The evolutionary pressure on males towards increased size is balanced by the penalty of an increased requirement for food. This double-edged aspect of size may explain the luxuriance of the male mane. The mane gives the appearance of great size without the drawbacks of increased weight. Confrontations between rival males are often settled before fighting takes place, the smaller of the two animals perceiving its disadvantage and withdrawing before coming to blows. The mane has other functions as well, such as protecting its owner against the claws and teeth of an opponent should fighting actually occur.

The bulk of a lion's diet comprises animals weighing 110–1,100lb (50–500kg), although it is an opportunistic feeder known to eat rodents, hares, small birds and reptiles. On the open plains, where cover is sparse, hunting primarily takes place at night but, where vegetation is thick, it may also occur during the day. Adult males rarely participate in hunts, probably because their mane makes them too conspicuous. When several lions stalk, they usually fan out and partially encircle the prey, cutting off potential escape routes. Although they can reach 36mph (58km/h) some of their prey can attain speeds of up to 50mph (80km/h), so lions must use stealth to approach to within about 100ft (30m) of their prey. From this distance they can

charge the prey and either grab or slap it on the flank before it outruns them. Lions do not take wind direction into account when hunting, even though they are much more successful when hunting upwind. Typically, only about one in four of lion charges ends successfully. Once knocked down, the prey has little chance of escape. Large animals are usually suffocated either by a bite to the throat or by clamping the muzzle shut.

The prey is usually eaten by all members of the group. When several lions feed together, or when the carcass is small, squabbles are frequent, but are usually brief and serious injuries are rare. Adult females require about 11lb (5kg) of meat per day and adult males 15.4lb (7kg).

Lions share their ranges with a variety of other carnivores such as leopards, cheetahs, wild dogs and spotted hyenas, each of which may feed on many of the same prey species as lions. But although all five species hunt animals weighing less than 220lb (100kg), only the lion regularly kills prey larger than about 550lb (250kg). Lions are also more likely to kill healthy adult prey than are the other carnivores. Hyenas are potentially the strongest competitors, being large-bodied noctural hunters. But by running down their prey, rather than stalking it as cats do, hyenas tend to kill calves and old and sick animals. Lions may actually benefit from the presence of hyenas, for in a study in the Ngorongoro Crater region of the Serengeti, in Tanzania, some 81 percent of all carcasses fed on by lions had been killed by hyenas.

Sexual maturity may be attained as early as 24–28 months in captivity and 36–46 months in the wild—a difference which may be due to nutritional factors. Females are sexually receptive more than once in a year, the receptive period lasting 2–4 days. The interval between cycles is highly irregular and may be between two weeks and several months. Ovulation is induced by copulation.

Gestation is short for such a large mammal—100–119 days. As a result, cubs are very small at birth and weigh less than one percent of the adult weight. Reproduction occurs throughout the year, although several females in a pride may give birth in the same month. Females rear their young together and will suckle cubs other than their own. Litter sizes vary from 1 to 5, with an average of 2 or 3. Cubs are weaned gradually and start eating meat at three months of age while continuing to nurse for up to six months from the female's four nipples. Mortality of cubs is high—as many as 80 percent may die before two years of

▲ **Lions mating.** While consorting, the adult male remains close to the receptive female TOP; other pride males may follow the pair from a distance. Either of the pair may initiate copulation by rubbing heads or by sniffing the other's groin. Copulation CENTER usually lasts about 20 seconds, during which the male usually emits a low growl and licks or bites the female's neck. The male usually dismounts abruptly as the female is likely to turn quickly and threaten with a snarl BELOW, or slap out. A pair may copulate up to 50 times in 24 hours.

◄ **Alert to any approach** – a group of lionesses, aroused while resting on a rocky outcrop in the heat of the day.

age. An adult female will produce her next litter when her cubs are two years old. If the entire litter dies, she will mate soon after the death of the last cub.

The lion is the most social of all the felids. Its social organization is based on the pride, which usually consists of 4–12 related adult females, their offspring and 1–6 adult males. Lions spend most of their time in one of several groups within the pride (see BELOW). Pride males may be related to each other, but are usually not related to the females. Both sexes defend the territory, although the males are more active in doing so. Territorial boundaries are maintained by roaring, urine marking and patrolling. Intruders usually withdraw at the approach of a resident, although males may fight and occasionally kill each other.

A pride will range over an area of between 8 and 155 sqmi (20–400 sqkm), depending on the size of the pride and the amount of game locally available. Large ranges may overlap with those of neighboring prides, although each pride has a central area for its exclusive use.

The varied environments in which lions can live profoundly affect their social behavior and ecology. Several studies of lions in Africa have shown that the lower the abundance of prey, the larger the territory must be. This relationship is clearest when measured during the season of lowest prey abundance. Factors such as rainfall, which govern the movements of prey, help to determine the severity of the lean season. The maximum size of a territory is deter-

▲ **Settled into a good meal,** three lions share a zebra carcass. As is usual, the actual kill was made by lionesses within their pride.

▶ **Straining every muscle,** a lion drags his kill to cover. This stray domestic horse was probably at a double disadvantage: on the one hand alone and ill adapted to its surroundings, and on the other made all the more conspicuous and attractive because of its differentness.

The Size of Groups Within the Pride

The members of a lion pride are normally scattered in several groups throughout the pride's range. The size of these groups—sometimes called "companionships" or "subprides"—is influenced by a number of ecological and social factors and is not merely a reflection of pride size: two prides, one of 7 and one of 20 animals, studied in Uganda, each had an average group size of 3 animals. Many factors favor the formation of larger groups: success in hunting large prey is increased and more kills can be stolen from other large carnivores. They also have fewer kills stolen by hyenas. One group of 2 lions studied had 20 percent of kills stolen by hyenas while larger groups of 6 or more lions lost only about 2 percent.

Other factors, however, restrict group size. If the available prey is small, few lions will be able to feed off one kill; and when prey is both scarce and small aggression among individuals at the kill increases, so that cubs

and juveniles may get little food. Lions feeding primarily on zebra can live in much larger groups than those killing small prey such as warthogs. Areas where prey is abundant throughout the year can support larger numbers of lions than areas where the supply of prey is low and irregular, but the sizes of groups found in each area are similar; there are simply more groups where prey is abundant. Whereas, for example, one area in Ruwenzori National Park supported some 35 tons of prey per sqmi (14,000 kg/sqkm) and one lion per sqmi (2.5 sqkm), another area with 7.05 tons per square mi (2,800 kg/sqkm) supported only one lion per 4.3 sqmi (11 sqkm). But because lions in each area fed on animals of about 220lb (100kg), the average group size in the two areas was identical. Apparently, while the average number of lions in a given area is affected by the relative abundance of prey, group size is affected little, if at all.

mined by the pride's ability to defend it and by the point at which social cohesion would otherwise break down.

Because they are poor competitors at kills, cubs can easily starve during their first year of life. Adult females may even prevent their own offspring from feeding during periods of food shortage. Even in times of abundant prey, cubs may die of starvation if only small animals are killed, because of the dominance of adults at the kill. By 18 months cubs are better able to secure food at kills, and by two years the survival of cubs is no longer related to the abundance of prey.

Instances of humans falling victim to lions are common. The Romans used lions imported from North Africa and Asia Minor as executioners, a practice which continued in Europe in medieval times. Attacks by man-eaters are also well known in the wild, although they are often perpetrated by injured or aged animals unable to kill their normal food; man is an easy prey, being neither swift nor strong. Many cases of man-eating have followed upon the extermination of the lions' normal supply of game. However, this has not always been the case. Towards the end of the last century, for example, two apparently healthy lions preyed regularly on the laborers of the Uganda–Kenya railway—so successfully that construction was halted.

While lions are not immediately threatened with extinction, their long-term survival is far from assured. In the past, local populations of lions were considerably reduced in numbers by hunters, who regularly killed up to a dozen per hunting trip. Today, hunting is regulated, but many lions are still killed illegally, trapped in snares set for other animals. A more significant threat comes from the fact that the game on which lions depend need large areas of land—a resource that is rapidly diminishing. As agriculture spreads, lions are quickly eliminated, either shot for their attacks on cattle or forced out as the game is destroyed.

KGVO

Blood Relatives

Kin selection in a lion pride

A lioness suckles the cubs of a female relative alongside her own; a male newcomer to the pride kills her cubs, but subsequently tolerates the boisterous play of the cubs he fathers. These and other unusual features of lion society can only be explained when it is known which lion is related to which.

Blood relationships among lions are discovered by keeping careful records of known individuals in a pride over a number of years. At the core of a lion pride are 4–12 related females. They are related because they grew up in that pride, as the offspring of related females. On average they are about as closely related as cousins. A pride probably persists for many generations, and if it grows larger than its optimum size, surplus subadult females ($2\frac{1}{2}$–3 years old) are driven out. These are not normally allowed to join other prides and as nomads will have a shortened life span and a reproductive success less than a quarter that of resident females.

Young subadult males are also driven out at $2\frac{1}{2}$–3 years old, if they do not leave of their own accord. They go as a group, with the other young males with whom they have grown up. Some of them may be brothers, littermates from the same lioness, but on average the adult males within a pride are about as closely related as half-brothers. Some are more distant relatives. The young male group remains together over the next year or two until, still as a group, they manage to take over as the breeding males of a pride. It is not likely to be the pride they grew up in, so they are not related to the females. Males may maintain tenure of a pride for periods as short as 18 months, or as long as 10 years, depending on the degree of competition from rival groups of males, and on the number of males in the coalition in possession.

A lioness will allow cubs which are not hers to suckle from her—cubs of four different mothers have been observed suckling at the same time from one lioness. This is most unusual among mammals—in most species a mother will not nurse offspring other than her own. The cubs which a lioness feeds, if they are not her own, are the offspring of her relatives. When she feeds any cubs, she is feeding young lions which carry a proportion of genes identical to her own. That proportion is a half if the cubs are her own offspring and the proportion is lower if the cubs are the offspring of a distant relative. But, in either case, by helping them with a supply of milk, she is helping to rear lions with some of her own genes.

Evolution has favored good parental behavior because that increases the number of the parents' genes which are passed on directly to future generations. Similarly, through the process known as kin selection, evolution also favors behavior which increases the number of an animal's genes which are passed on indirectly, via the offspring of relatives. This does not imply, or require, the tolerant lioness to be conscious of kin selection; evolution has merely made her behave tolerantly towards her pride companions' offspring because they are related to her.

The males in a pride are surprisingly close companions: they fight fiercely and cooperatively against strange males, but they do not fight each other for receptive females. Instead they operate a kind of gentleman's agreement whereby the first male to encounter a female in heat is usually accepted as being dominant over other males.

▶ **Lion pride at rest,** showing the relationships to the lioness marked (1). She is suckled by three cubs, one her own (i), the others the offspring of the two females 3 and 6. (2) Her half sister. (3) Her first cousin who has two cubs (iii). (4) Her second cousin. (5) Her elderly mother. (6) Her daughter who has two cubs (vi). (7) Her daughter who has three cubs (vii). (8), (9) Adult males who are half brothers to each other but not related to any of the pride females.

▷ **Lions and cubs.** An adult lion ABOVE with a cub he has just killed. Whenever a coalition of males takes over a pride they are liable to kill cubs sired by the ousted males. The new pride males soon have their own offspring to which they are extremely tolerant BELOW.

▼ **Close companions,** these males are related as are most pride males. Social bonds formed during grooming such as this show clear benefits when the males have to make a coordinated defense of the pride against intruding males.

Lions have good reason not to fight in such cases. Firstly, the chances are very low of any one mating resulting in a reared cub. Secondly, and more important, if a male lets his related companion mate instead, some of his own genes are nevertheless still passed on to any cubs fathered.

It is also not in the lion's long-term interest to fight with his companions, because a male needs companions to defeat the rival groups of males waiting to take over his pride. The biggest groups of males which take over a pride, those of 4–6, manage to keep possession of prides for 4–8 years, much longer than pairs can. The teamwork needed is possible only among companions who do not quarrel.

An established adult male in a pride is usually friendly towards the females and toward the cubs fathered by him or by his related companions. A member of a newly arrived male group behaves very differently. He is liable to kill at least some of the cubs in the pride when he takes it over. This violent and apparently unadaptive behavior was at first puzzling—most mammals do not ordinarily kill the young of their own species. However, from records of the life histories of lions in prides over several years, it is now clear that if males kill cubs when they take over prides, they probably leave more descendants of their own. A male is not related to the cubs he kills, but by killing them he can make their mothers produce his own offspring sooner (by becoming receptive to him soon after the death of her last infant). His cubs will also survive better if there are no older competing cubs present. Thus, killing cubs in such circumstances is adaptive and, like other lion behavior, is an aspect of the process of kin selection at work.

BCRB

TIGER

Panthera tigris [E]
One of 5 species of the genus *Panthera*.
Family: Felidae.
Distribution: India, Manchuria, China,
Indonesia.

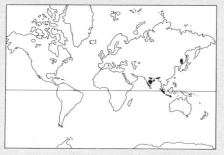

Habitat: Varied, including tropical rain forest,
snow-covered coniferous and deciduous forests,
mangrove swamps and drier forest types.

Size: Male Indian: head-to-
tail-tip 8.8–10.2ft
(2.7–3.1m); shoulder height
3ft (91cm); weight
397–573lb (180–260kg);
female: head-to-tail-tip
7.8–9.4ft (2.4–2.8m);
weight 287–353lb
(130–160kg). Male Javan
and Sumatran: head-to-tail-
tip 7.2–8.9ft (2.2–2.7m);
weight 220–330lb
(100–150kg).

Gestation: 103 days.

Longevity: about 15 years (to 20 in captivity).

Subspecies: 8 (see p30).

[E] Endangered

FEW animals evoke such strong feelings of fear and awe as the tiger. For centuries its behavior has inspired legends, and the occasional inclusion of man in its diet has intensified the mystique.

Tigers are the largest living felids. Siberian tigers are the largest and most massively built subspecies; the record was a male weighing 846lb (384kg).

Like that of other big cats, the tiger's physique reflects adaptations for the capture and killing of large prey. Their hindlimbs are longer than the forelimbs as an adaptation for jumping; their forelimbs and shoulders are heavily muscled—much more than the hindlimbs—and the forepaws are equipped with long, sharp retractile claws, enabling them to grab and hold prey once contact is made. The skull is foreshortened, thus increasing the shearing leverage of the powerful jaws. A killing bite is swiftly delivered by the long, somewhat flattened canines.

Unlike the cheetah and lion, the tiger is not found in open habitats. Its niche is essentially that of a large, solitary stalk-and-ambush hunter which exploits medium- to large-sized prey inhabiting moderately dense cover (see p30).

The basic social unit in the tiger is mother and young. Tigers have, however, been successfully maintained in pairs or groups in zoos and are seen in groups (normally a female and young, but sometimes a male and female) at bait kills in the wild, indicating a high degree of social tolerance. The demands of the habitat in which the tiger lives have not favored the development of a complex society and instead we see a dispersed social system. This arrangement is well suited to the task of finding and securing food in an essentially closed habitat where the scattered prey is solitary or in small groups. Under these circumstances, a predator gains little by hunting cooperatively, but can operate more efficiently by hunting alone.

In a long-term study of tigers in Royal Chitawan National Park, in southern Nepal, it was found, using radio-tracking techniques, that both males and females occupied home ranges that did not overlap those of others of their sex; home ranges of females measured approximately 8sqmi (20sqkm) while males had much larger ones, measuring 23–40sqmi (60–100sqkm). Each resident male's range encompassed those of several females. Transient animals occasionally moved through the ranges of residents, but never remained there for long. By comparison, in the Soviet Far East, where the prey is scattered and makes large seasonal movements, the density of tigers is low, less than one adult per 40sqmi (100sqkm).

Tigers employ a variety of methods to maintain exclusive rights to their home range. Urine, mixed with anal gland secretions, is sprayed onto trees, bushes and rocks along trails, and feces and scrapes are left in conspicuous places throughout the area. Scratching trees may also serve to signpost. These chemical and visual signals convey much information to neighboring animals, which probably come to know each other by smell. Males can learn the reproductive condition of females, and intruding animals are informed of the resident's presence, thus reducing the possibility of direct physical conflict and injury, which the solitary tiger cannot afford as it depends on its own physical health to obtain food. The importance of marking was evident in the Nepal study, when tigers which failed to visit a portion of their home range to deposit these "occupied" signals (either due to death or confinement with young) lost the area in three to four weeks to neighboring animals. This indicates that boundaries are continually probed and checked and that tigers occupying adjacent ranges are very much aware of each other's presence.

The long-term exclusive use of a home range confers considerable advantages on the occupant. For a female, familiarity with an area is important, as she must kill prey

▲ **A tiger drinks frequently** during a meal, and in the wild will often drag its dead prey into cover in the vicinity of water.

◄ **The striking "white tiger"** was once not unusual in north and east central India, where the forbear of this zoo-bred animal originated.

► **Land tenure systems** of tigers in Chitawan National Park, Nepal. Each male's range encompasses those of several females. There is little or no overlap between individuals of the same sex in Chitawan. In other places female ranges may overlap.

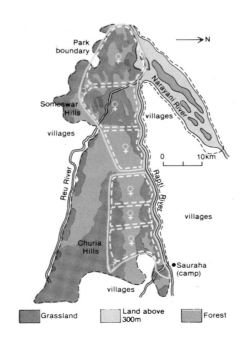

with some regularity to raise young. When the young are small and unable to follow she must obtain food from a small area, as she has to return to suckle them at regular intervals. Later, when her young are larger and growing rapidly she must be able to find and kill enough prey to feed herself and the young.

Territorial advantages for males seem to be different; they maintain ranges three or four times larger than those of females, so food is not likely to be the critical factor. What matters is access to females and paternity of cubs. Males are not directly involved in the rearing of young. Although there is not as much evidence as for lions (see p27), several instances have been reported of male tigers killing cubs. These are usually associated with the acquisition of one male's home range by another. By killing the offspring of the previous male, the incoming male ensures that females in his newly acquired range come into heat and bear his offspring.

Tigers living in areas of prime habitat raise more young than can find openings, so large numbers of animals, usually young adults, live on the periphery. There is no clear picture of the social organization in these marginal areas, but ranges are certainly larger and probably overlapping, and there is little successful reproduction.

This outlying segment of the population is important, as it promotes genetic mixing in the breeding population and ensures that there are enough individuals to fill any vacancies that may arise. Unfortunately, it is usually these tigers that come into conflict with humans, as the habitat they occupy is, more often than not, heavily exploited by man and his livestock.

Sexual maturity is reached by 3–4 years of age. Breeding activity has been recorded in every month for tigers from tropical regions, while in the north breeding is restricted to the winter months. A female is only receptive for a few days and mating may take place as many as 100 times over a period of two days. Three to four cubs, weighing about 2.2lb (1kg) each, are born blind and helpless. The female rears them alone, returning to the "den" site to feed them until they are old enough to begin following her, at about eight weeks of age. The cubs remain totally dependent on their mother for food until they are approximately 18 months old and may continue to use their mother's range until they are 2–2½ years old, when they disperse to seek their own home ranges.

All the surviving subspecies are endangered. Its broad geographical distribution, which encompasses such a variety of habitat types, creates the illusion that the tiger is an adaptable species. In fact, it is a highly specialized large predator with very specific ecological requirements and is much less adaptable than, say, the leopard. Once found across much of Asia, the tiger's present distribution and reduced numbers indicate that the requirements for large prey and sufficient cover are becoming more difficult to meet as areas suitable for large wild hoofed mammals, and consequently tigers, are being appropriated for agricultural purposes. As most tiger reserves are relatively small, less than 390sqmi (1,000sqkm), and isolated, the effective population-size is small and there is little or no inter-breeding between populations.

Tigers only rarely become man-eaters; indeed they normally avoid contact with man. Some man-eaters may be old or disab-

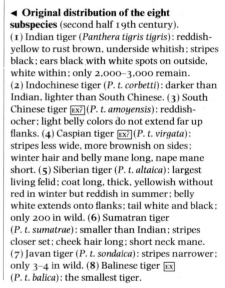

1

2

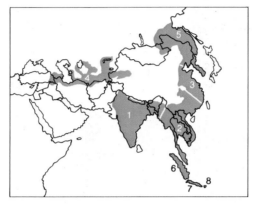

led but there are also many cases of healthy, young adult tigers acquiring the habit. This behavior may begin with an accident—a sudden close encounter that ends with the person being killed. Sometimes a single episode may be all that is required for a tiger to learn to kill a man. Whether or not a tiger takes the next step and becomes a deliberate man-eater may depend on the opportunity. There is also some suggestion that "aversive" encounters with people over the first human kill may discourage further incidents. The availability of other prey may also be a factor.   MS

◀ **Original distribution of the eight subspecies** (second half 19th century). (1) Indian tiger (*Panthera tigris tigris*): reddish-yellow to rust brown, underside whitish; stripes black; ears black with white spots on outside, white within; only 2,000–3,000 remain. (2) Indochinese tiger (*P. t. corbetti*): darker than Indian, lighter than South Chinese. (3) South Chinese tiger [EX?] (*P. t. amoyensis*): reddish-ocher; light belly colors do not extend far up flanks. (4) Caspian tiger [EX?] (*P. t. virgata*): stripes less wide, more brownish on sides; winter hair and belly mane long, nape mane short. (5) Siberian tiger (*P. t. altaica*): largest living felid; coat long, thick, yellowish without red in winter but reddish in summer; belly white extends onto flanks; tail white and black; only 200 in wild. (6) Sumatran tiger (*P. t. sumatrae*): smaller than Indian; stripes closer set; cheek hair long; short neck mane. (7) Javan tiger (*P. t. sondaica*): stripes narrower; only 3–4 in wild. (8) Balinese tiger [EX] (*P. t. balica*): the smallest tiger.

Hunting Technique

▲ **Sumatran tiger in stream.** During the hot season tigers spend much of the daytime resting near streams or other water courses and often lie or stand in water to keep cool.

◄ **Camouflage of tigers.** A tigress in tall grass illustrates the advantage of the cryptic coat coloration. The stripes disrupt the outline of the body as the hunter stalks or lies in ambush for its prey.

Tigers hunt alone, actively searching for prey more often than waiting in ambush. An individual will typically travel 6–12mi (10–20km) during a night of hunting. Tigers do not easily catch their prey—probably only one in 10 or 20 tries is successful.

Having located the quarry, a stalking tiger then uses sight. The tiger makes maximum use of cover for concealment to move closer to the prey (1). It must approach to within 66ft (20m) or less if the final rush is to be successful. The approach is extremely cautious, with the tiger placing each foot carefully on the ground and pausing from time to time to assess the situation. It assumes a semi-crouch or crouch, with the head up, during the stalk. Having made use of the distance and position of the prey, the tiger gathers itself up and suddenly rushes its victim (2), covering the intervening distance in a few bounds. When contact is made, the momentum of the charge may knock the animal off its feet, or if the prey is in flight a slap with a forepaw may serve to throw it off balance. A tiger's attack is usually from the side or from the rear; it does not launch itself into the air or spring on its prey from a distance. While it is seizing the prey about the shoulder, back or neck (3) with its claws, the tiger's hind feet usually do not leave the ground. At this point, the prey is jerked off its feet, if it hasn't happened earlier in the attack. A bite to the throat or neck may be delivered upon contact or while the tiger brings the victim to the ground (4).

When the prey weighs more than half as much as the tiger, the throat bite is commonly used and death is most likely caused by suffocation. (See p37 for description of physical adaptations to hunting large prey.) The grip may be retained for several minutes after death. Kills are carried or dragged into dense cover and tigers usually commence feeding on the rump. It is not unusual for a tiger to consume 44–77lb (20–35kg) of meat in a night, but the average eaten over several days is less, about 33–40lb/day (15–18kg).

Tigers stay near their kill and continue to feed at their leisure until only skin and bones remain—the average time in the Chitawan National Park was three days at each kill. Small prey, such as Barking deer, are eaten in one meal, whereas the larger sambar, elk and bison provide food for several days unless several tigers (usually females and young) are feeding on the carcass.

A tigress with young has to kill more often to provide food—an estimated once every 5–6 days, or 60–70 animals per year, for a female with two young. This compares with a kill every 8 days or 40–50 kills per year for a female in the same area without dependent young.

A tiger will eat whatever it can catch, but the larger hoofed mammals (prime adults, as well as young or aged animals) in the 110–440lb (50–200kg) range form the bulk of their diet. Typical prey are thus sambar, chital, Swamp deer, Red deer, Rusa deer and Wild pigs. Tigers occasionally take very large prey such as rhino and elephant calves, water buffalo, moose, wapiti and gaur. In many areas, agricultural stock are also readily taken, especially where wild prey is depleted.

CHEETAH

Acinonyx jubatus [V]
Sole member of genus.
Family: Felidae.
Distribution: Africa, S Asia, Middle East.

Habitat: most habitats in Africa except rain forest.

Size: head-body length 44–53in (112–135cm); tail length 26–33in (66–84cm); weight 86–143lb (39–65kg). Males usually slightly larger than females.

Coat: tawny with small round black spots. Face marked by conspicuous "tear stripes" running from the corner of the eyes down sides of nose; cubs under three months old blackish, with a mantle of long blue-gray hair on top of the back and neck.

Gestation: 91–95 days.

Longevity: up to 12 years (17 in captivity).

Subspecies: 2 **African cheetah** (*A. j. jubatus*) and **Asiatic cheetah** [E] (*A. j. venaticus*). (**King cheetah**, a mutant form occuring only in S Africa, was once incorrectly described as a separate species, *Acinonyx rex*. Coat: spots along spine joined together in stripes, with small splotches on the body.)

[E] Endangered. [V] Vulnerable.

THE fastest animal on land, the cheetah can sprint at up to 60 miles an hour (96km/h) for a brief part of its chase. It still occurs over most of Africa, but very few now remain in southern Asia (where it probably evolved) and the Middle East.

The cheetah is easily distinguished from other cats, not only by its distinctive markings, but also by its loose and rangy build, small head, high-set eyes and small, rather flattened ears. The usual prey consists of gazelles, impala, wildebeest calves and other hoofed mammals up to 88lb (40kg) in weight. In some areas, hares are also an important food. The prey is hunted by stalking from a few seconds up to several hours, until the prey is within 100ft (30m). before chasing. About half the chases are successful and an average chase is 550ft (170m) and lasts 20 seconds, rarely exceeding one minute. The prey is suffocated by biting the underside of the throat. On average, an adult eats 6.2lb (2.8kg) of meat per day. Drinking is seldom more frequent than once every four days and sometimes as infrequent as once in 10 days.

Sexual maturity occurs at 20–23 months old. Courting females and males probably already know each other because their home ranges overlap. Females in heat squirt urine on bushes, tree trunks and rocks to attract males which, when they discover the scent, hurriedly follow the trail, calling with yelps. The receptive female responds to the yelps by approaching the male. Mating sometimes occurs immediately, with copulation lasting less than one minute. They stay together for a day or two and mate several times. The males have a hierarchy and apparently it is usually only the dominant male that mates, while his companions wait nearby.

There is no regular breeding season and cubs are born in all months. The litter size is 1–8, but the average is three. Newborn cubs weigh 8–11oz (250–300g) and are up to 12in (30cm) long from the nose to the root of the tail. Their eyes open at 2–11 days old. Cubs remain hidden under bushes or in dense grass, but their mother carries them to a new hiding place every few days. By 5–6 weeks old, cubs are able to follow their mother and begin eating from the prey their mother catches. Males do not help to raise

▲ **A rate of acceleration** comparable to that of a high-powered sports car enables the cheetah to outrun all other animals over short distances.

▲ **A solitary hunter,** the cheetah uses a stalk-and-rapid-chase technique. This female has spotted her quarry and is about to begin stalking.

▶ **A strangling throat bite** killed this Thomson's gazelle; the carcass is now dragged away to cover.

◀ **Mantle of blue-gray hair** indicates that these cubs, feeding with their mother, are under three months old.

The Cheetah's Niche

Where cheetahs are found so also are other large carnivores such as lions, leopards, hyenas, wild dogs and jackals—and other meat-eaters such as vultures. But if different species are to coexist in the same area they must exploit available resources in ways that minimize the likelihood of direct competition and open conflict. One way of achieving this is to evolve an anatomy that is highly specialized for a particular method of hunting.

A slender build and highly flexible spine enable the cheetah to make astonishingly long and rapid strides; and, unlike other cats, the cheetah's claws when retracted are not covered by a sheath but are left exposed to provide additional traction during rapid acceleration. However, with great sprinting prowess comes limited endurance and this means that the cheetah can only hunt effectively in open country where there is enough natural cover for stalking.

A sure method of killing prey is also important. The small upper canine teeth have correspondingly small roots bounding the sides of the nasal passages, permitting an increased air intake that enables the cheetah to maintain a relentless suffocating bite.

The cheetah usually hunts and eats later in the morning and earlier in the afternoon than other large carnivores, which tend to sleep in the heat of the day; its less-developed whiskers suggest less nocturnal activity than other cats.

Greater daytime activity, however, brings the cheetah into contention with vultures—soaring on daytime thermals. Vultures sometimes drive a cheetah away from its kill and their descent also attracts other carnivores who may then appropriate the cheetah's meal. The problem is minimized by the cheetah's stealth as a hunter and by its habit of dragging its prey to a hiding place before eating.

the cubs. Weaning occurs at about three months of age. Fewer than one-third of the cubs, on average, survive to adulthood.

Adult females are solitary, except when they are raising cubs. They rarely associate with other adults, and when they do, it is likely to be for only a few hours following a chance encounter with a sister or when found by territorial males. Males are more gregarious than females and often live in permanent groups, which are sometimes composed of littermates.

In the 16th century, cheetahs were commonly kept by Arabs, Abyssinians and the Mogul emperors to hunt antelopes. More recently, cheetahs have been in demand for their fur, which is used for women's coats. In the wild, cheetahs are widely protected, but so long as the trade in skins in many European countries and Japan remains legal, widespread poaching will continue to occur. An estimated 5,000 cheetah skins were traded annually in recent years.

A more substantial threat to the cheetah's survival is the loss of habitat, which deprives it of suitable prey, reduces its hunting success, causes more cubs to die of starvation and fall victim to predators, increases the proportion of kills stolen by other large carnivores, and causes conflict with man through increased attacks on domestic livestock. Captive breeding, although successful, is not a suitable alternative to preserving the natural habitat. The total surviving cheetah population in Africa is probably only about 25,000.   GWF

Cheetahs of the Serengeti
Male–female differences in habitat exploitation

Spacing behavior among cheetahs shows how they exploit their habitat and prey. In areas such as the Serengeti in East Africa, where prey species are migratory, adult female cheetahs (with or without cubs) migrate annually over a home range of about 310sqmi (800sqkm). Each adult female travels her home range in an annual cycle and appears to use the same area year after year.

Cheetah litters separate from their mother when they are adult size, at 13–20 months old. The siblings usually remain together for several months longer. One by one the females, when 17–23 months old, leave their littermates.

Female cheetahs, although not territorial, avoid each other. Adult females are not aggressive to other females or to males, but if they see another cheetah nearby they usually walk farther away or hide. Their mutual avoidance means that non-related or distantly related females, as well as close relatives, have home ranges that overlap each other, but in which they rarely interact.

Young adult male cheetahs leave their mother's home range as a group. Apparently they are chased away by older and stronger territorial males. The young males disperse about 12mi (20km), and probably sometimes much farther, beyond their mother's home range. Adult male littermates often remain together for life and non-littermates sometimes join together in groups of 2–4.

Territorial males defend a well-defined area throughout which they regularly mark prominent trees, bushes and rocks with urine, feces and scratch marks. Territories of males in the Serengeti cover about 12sqmi

▲ **A vigilant female** sits aloft on a termite mound for a better view of her surroundings, while her cubs frolic nearby.

◄ **Scent-marking its territory,** a male cheetah sprays urine backward onto a conspicuous landmark, one of many similarly marked by the same animal in its territory.

► **In defense of their territory,** males close in on one of three intruding males; this one was subsequently killed. It is now known that male cheetahs, often close relatives, sometimes work together to hold and defend a territory. Previously it was supposed that a territory was always held and defended by one dominant animal.

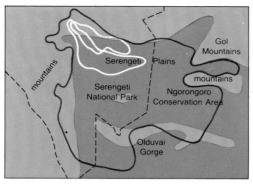

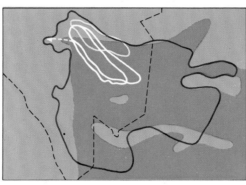

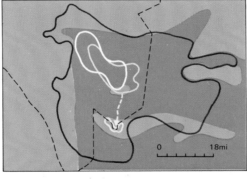

▶ **Home ranges** of members of three cheetah families observed on the Serengeti Plains, Tanzania, over the same period. Home ranges of other families overlapped these. TOP The home ranges of two female cheetah littermates partly overlap each other and the ranges (not shown) of their mother, other females, territorial males and nomadic males. The home ranges of these sisters are large because they follow the migratory prey. Each home range shown represents the limit of movement in a full year: for several weeks, the cheetah remains in one locality within her range, making zigzag and circular movements in search of prey. When hunting becomes poor, she moves on a few kilometers to a new locality.

CENTER In this typical cheetah family two young adult daughters (white) remain near and overlap their mother's home range (blue), while the young adult males (dotted yellow line) probably left because of aggression by territorial males in their mother's area, and they remain nomadic until they are able to defend a territory. About half the males remain in groups, whereas females are always solitary. Some male groups consist of littermates, some consist of males who were born to different mothers and some groups are a mixture of both. Male group size is 2–4.

BOTTOM The home ranges of this mother (blue) and daughter (white) are entirely in the grasslands. Cover is, however, available in drainages with tall dense herbs and in rocky outcrops with bushes. In this case the two sons (dotted yellow line) emigrated more than 11mi (18km) from their mother's home range, ousted two territorial males from a woodland territory and established themselves there.

woodland   - - - - boundary of Serengeti National Park

grassland   ——— boundary of study area

(30sqkm). Territorial males do not migrate 30–50mi (50–80km) to follow the prey, as the females do, but when there is no food or water within the territory they temporarily leave to feed and drink nearby. Lone males and groups of males are known to hold their territories for at least four years, but eventually they are ousted or killed by stronger males, either another lone male or a group of males in coalition.

The males' tendency to live in small groups, as well as to hunt and eat together, is most probably due to an increased success in establishing and defending a territory compared with the chance they would have as solitary males. The males hold territories in places of moderate vegetative cover, such as woodlands and bushed drainages (see movements of emigrating males, LEFT).

Not all male cheetahs are territorial; some males seem to be nomadic. These nomads frequently encounter territorial males, who respond aggressively to them. One fight observed between a group of three territorial males and a group of three intruding males began when the territorial males chased and caught one of the intruders. All three territorial males fought with the intruder, biting him repeatedly all over his body and pulling out mouthfuls of fur. Eventually one of the territorial males inflicted a suffocating bite on the underside of the neck, the same bite that is used in killing prey.

Immediately after killing the first intruder, the territorial males walked towards the other two intruders, who were watching from about 300yd (275m) away. They fought briefly, then one territorial male chased an intruder at least 1,100yd (1km). Soon after, all three fought with the remaining intruder, but eventually left him alone. The result of this territorial encounter was one intruder lying dead, one injured and one chased away. The defenders were unharmed except for one bloody lip.

The social system of male territoriality in itself restricts the density of cheetahs. When the cheetah population increases, more of the available habitat is claimed by territorial males, leading to increased conflict and more deaths. Females, too, are affected through increased harassment from the sexually motivated males. Sometimes, territorial males intent on mating virtually hold a mother cheetah captive for a day or two, which prevents her from tending her cubs. This probably leads to a greater number of cub deaths, by making cubs more conspicuous in their behavior and therefore vulnerable to predators, and by reducing the mother's ability to feed them.   GWF

LEOPARD

Panthera pardus [V]
One of 5 species of the genus *Panthera*.
Family: Felidae.
Distribution: Africa S of the Sahara, and S Asia; scattered populations in N Africa, Arabia, Far East.

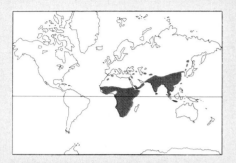

Habitat: most areas having a reasonable amount of cover, a supply of prey animals and freedom from excessive persecution; from tropical rain forest to arid savanna; from cold mountains almost to urban suburbs.

Size: head-body length 40–75in (100–190cm); tail length 28–37in (70–95cm); shoulder height 18–32in (45–80cm); weight 66–155lb (30–70kg). Males are about 50 percent larger than females.

Coat: highly variable, essentially black spots on a fawn to pale brown background. Typically, the spots are small on the head, larger on the belly and limbs, and arranged in rosette patterns on the back, flanks and upper limbs.

Gestation: 90–105 days.

Longevity: up to 12 years (20 in captivity).

Subspecies: 7. **Amur leopard** [E] (*P. p. orientalis*), Amur-Ussuri region, N China, Korea; coat long and thick, light-hued in winter, reddish-yellow in summer; spots large. **Anatolian leopard** [E] (*P. p. tulliana*), Asia Minor; coat brighter and tanner, often with some gray hues. **Barbary leopard** [E] (*P. p. panthera*), Morocco, Algeria, Tunisia. **North African leopard** (*P. p. pardus*), Africa except extreme N, Asia; coat yellowish-ocher. **Sinai leopard** [E] (*P. p. jarvis*), Sinai; coat light with large spots. **South Arabian leopard** [E] (*P. p. nimr*). **Zanzibar leopard** [EX?] (*P. p. adersi*), Zanzibar; spots very small.

[E] Endangered.　[EX?] Probably extinct.　[V] Vulnerable.

► **A North African leopard** licking a paw clean, rather like a big domestic cat. The coat spots provide excellent camouflage, especially in trees; and the prominent, extremely sensitive whiskers are those of an animal that hunts at night.

CONFUSION surrounds leopards. Even the name "leopard" originates from the mistaken belief that the animal was a hybrid between the lion (*Leo*) and the "pard" or panther. And only just over a century ago it was still disputed whether leopards and "panthers" were separate species. In fact the words "pard" and "panther" are vague, archaic terms that have been used for several large cats, especially the leopard, jaguar and puma. With luck, the confusing term "panther" will die out before the single species, best called the leopard, does.

About 30 subspecies have been named, but only about seven are still accepted today. The commonest form is the North African leopard, which occurs over most of the leopard's range. The other subspecies are small or geographically isolated populations.

In form the leopard is average among the large cats—slender and delicate compared with the jaguar, but sturdy and stolid compared with the cheetah. There are various aberrant coat patterns. One of the commonest and most striking is melanism, the leopard being totally black. It is caused by a recessive gene, which is apparently more frequent in leopard populations in forests, in mountains and in Asia. In the Malay peninsula as many as 50 percent of leopards may be black; elsewhere the proportions are much lower. The name "Black panther" is sometimes erroneously applied to such animals in the belief that they are a distinct species. Several other cat species, including the jaguar and serval, also exhibit melanism.

The leopard is the most widespread member of the cat family, and this is largely due to its highly adaptable hunting and feeding behavior. Leopards catch a great variety of small prey species—mainly small mammals and birds—and they do so by a combination of opportunism, stealth and speed. They hunt alone, generally at night, and either ambush their prey or stalk to within close range before making a short fast rush. Adept tree climbers, leopards often drag their prey up trees, out of reach of scavengers. Because of the variety and small size of their prey, leopards avoid strong competition with such carnivores as lions, tigers, hyenas and African wild dogs, which depend on larger prey.

Over most of their range, leopards have no particular breeding season. Females are sexually receptive at 3–7 week intervals, and the period of receptivity lasts for a few days, during which mating is frequent. Most litters consist of usually three (range 1–6) blind, furred cubs weighing 15–20oz (430–570g).

The cubs are kept hidden until they start to follow the mother at 6–8 weeks old. Only the mother cares for the young. She does so until her cubs are about 18–20 months old, whereupon she mates again. Sexual maturity is probably achieved in leopards at about $2\frac{1}{2}$ years.

▲ **The legendary "black panther"** was once thought to be a distinct species. It is now known to be a black-coated form of the leopard. The way the light falls on this individual clearly reveals the familiar leopard spots against the black fur.

HOW LEOPARDS USE TREES

◄ **Resting.** Safely aloft and shaded from the midday sun, a leopard dozes in the branches of an acacia tree.

▼ **Hunting.** Ready to leap, a leopard surveys its surroundings for potential prey. Occasionally a leopard will drop directly onto a passing animal from the cover of a tree.

► **Conserving food.** Leopards often drag their kills – in this case a topi calf – up trees, where they can eat and store them out of reach of most scavengers.

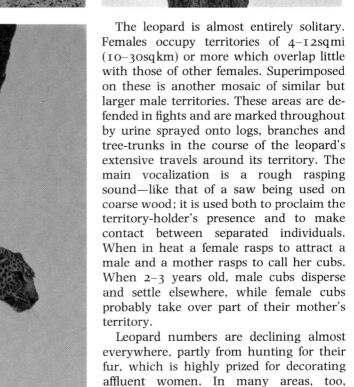

The leopard is almost entirely solitary. Females occupy territories of 4–12sqmi (10–30sqkm) or more which overlap little with those of other females. Superimposed on these is another mosaic of similar but larger male territories. These areas are defended in fights and are marked throughout by urine sprayed onto logs, branches and tree-trunks in the course of the leopard's extensive travels around its territory. The main vocalization is a rough rasping sound—like that of a saw being used on coarse wood; it is used both to proclaim the territory-holder's presence and to make contact between separated individuals. When in heat a female rasps to attract a male and a mother rasps to call her cubs. When 2–3 years old, male cubs disperse and settle elsewhere, while female cubs probably take over part of their mother's territory.

Leopard numbers are declining almost everywhere, partly from hunting for their fur, which is highly prized for decorating affluent women. In many areas, too, leopards are persecuted because of their attacks on domestic livestock. Numbers will continue to decline, but despite the likely loss of some subspecies, the adaptable leopard will probably continue to thrive in many areas where human population pressures are low. There are still well over 100,000 left.

Leopards are a highly popular attraction for visitors to National Parks. Elsewhere, relations with man are mutually hostile. In Asia, very occasionally, leopards become man-eaters and individuals have been known to kill over one hundred people. As well as killing them for profit and to reduce loss of livestock, man also kills leopards for sport—in Africa the leopard is one of the "Big Five" most highly rated prey of the Western sport hunter, the other favored species being the lion, buffalo, elephant and rhinoceros.   BCRB

OTHER BIG CATS

Three species in 2 genera
Family: Felidae.

Jaguar   Snow   Clouded
leopard   leopard

▶ **The rare Snow leopard** patrols large territories in its remote Asian homeland. A superb jumper, the Snow leopard may ascend above 19,000ft (6,000m) in summer in pursuit of prey.

▼ **The jaguar climbs well** but usually stalks its prey on the ground. Larger than the Old World leopard it somewhat resembles, the jaguar has a more compact body, more reddish coloration, a broader head and more powerful paws.

THE jaguar, Snow leopard and Clouded leopard occur in quite different regions of the world, but all live in mostly forested wilderness habitats. Their numbers are low and diminishing, partly as a result of the demand for their attractive pelts by the fur trade—a practice that is now banned. All three species are illustrated on pp18–19.

The **jaguar** is the only member of the genus *Panthera* (big cats) to be found in the Americas, where it is considered to be the New World equivalent of the leopard.

Although the jaguar is classified with the big cats, which can roar, it does not seem to do so, a characteristic it shares with the Snow leopard. It grunts frequently when hunting and will snarl or growl if threatened. The male also has a mewing cry used in the mating season. The jaguar has a compact body with a large broad head and powerful paws.

Jaguars prefer dense forest or swamps with good cover and easy access to water, although they will hunt in more open country if necessary. They swim and climb very well, but usually stalk prey on the ground. Prey species include peccary, deer, monkeys, tapir, sloths, agouti, capybara, birds, caymen, turtles, turtle eggs, frogs, fish and small rodents. They will also take domestic stock if it is easily available. The prey is often cached by burying.

It is quite widely believed, particularly among the Amazonian Indians, that jaguars catch fish which they deliberately lure to the surface by twitching their tail in the water, flicking the fish onto the bank with a forepaw. It seems more likely that as the jaguar crouches in ambush on the bank, its tail occasionally hits the surface of the water, by which it may inadvertently attract fish.

Jaguars are solitary, except during the breeding season, and maintain a territory which varies from 2–200sq mi (5–500sq km), depending on prey density. They are occasionally known to travel up to 500mi (800km) but why they undertake such a journey is unknown.

Two to four young, each weighing 25–32oz (700–900g), are born at a time. They are blind at birth, but open their eyes after about 13 days and remain with their mother for two years. Sexual maturity is achieved at three years.

The shy, nocturnal and virtually unknown **Snow leopard** or **ounce** is classified with the big cats, but shares some small cat characteristics, for example it does not roar and it feeds in a crouched position.

The Snow leopard has to contend with extremes of climate and its coat varies from fine in summer to thick in winter. The surfaces of its paws are covered by a cushion of hair which increases the surface area, thus distributing the animal's weight more evenly over soft snow and protecting its soles from the cold.

Prey density is usually very low and territories are therefore large, probably up to 40sq mi (100sq km). Snow leopards move to different altitudes along with migrating prey, which include ibex, markhor, wild sheep, Musk deer, as well as marmots, Piping hare, bobak, tahr, mice and birds; in winter, deer, wild boar, gazelles and hares form a major part of their diet. Snow leopards usually stalk their prey, springing upon it, often from 20–50ft (6–15m) away.

Snow leopards are solitary except during the breeding season (January to May), when male and female hunt together, or when a female has young. One to four young are born in spring or early summer in a well-concealed den lined with the mother's fur. Initially, the spots are completely black. The young open their eyes at 7–9 days, are quite active by two months and remain with their mother through their first winter.

Snow leopards are extremely rare in many parts of their range due to the demand for their skins by the fur trade. Although in many countries it is now illegal to use these furs, the trade continues and the species remains under threat.

Neither truly a big cat nor a small cat, the **Clouded leopard** provides a bridge between the genera *Panthera* (lion, tiger etc) and *Acinonyx* (cheetah) on the one hand and the genus *Felis* (small cats) on the other, sharing

characteristics of both groups. It differs from the big cats in having a rigid hyoid bone in its vocal apparatus, which prevents it roaring, and from the small cats by the low level of grooming and by its posture when at rest, lying with its tail directed straight behind and its forelegs outstretched.

The Clouded leopard is heavily built, with short legs and a long tail. It has the usual felid complement of 30 teeth but the upper canines are relatively long and, in conjunction with the incisors, are used to tear meat from prey as the leopard jerks its head upwards. The snout is rather broad, although the head is quite narrow. The retina is most often yellow and the pupils contract to a spindle.

The Clouded leopard is an arboreal cat preying mainly upon monkeys, squirrels and birds, which it often swats with its broad, spoon-shaped paws. It is an adept climber and can run down trees head-first, clamber upside down on the underside of branches and swing by a single hindpaw before dropping directly onto deer or wild boar, which are its main terrestrial prey. The Clouded leopard is active at twilight, resting and sleeping in the treetops for the rest of the day and night.

Two to four blind and helpless young are born; their coloration and coat patterning differs from the adult's in that the large spots on the sides are completely dark. Cubs open their eyes after 10–12 days and are quite active by five weeks. The young Clouded leopard probably achieves independence by nine months.

Clouded leopards are elusive, and there is no information about the social behavior of this species in the wild. GK

Abbreviations: HBL = head-body length; TL = tail length; wt = weight.
V Vulnerable.   E Endangered.

Jaguar V
Panthera onca

SW USA to C Patagonia. In tropical forest, swamps and open country, including desert and savanna. HBL 44–73in; TL 18–29in; shoulder height 27–30in; wt 126–249lb. (Females on average 20% smaller.) Coat: basically yellowish-brown but varying from almost white to black, with a pale chest and irregularly placed black spots on belly; back marked with dark rosettes; lower part of tail ringed with black; a black mark on the lower jaw near the mouth; outer surface of ear pinnae black. Gestation: 93–110 days. Longevity: up to 20 years in captivity. Subspecies: 8. **Yucatán jaguar**

(*P. o. goldmani*), SW Yucatán (Mexico), N Guatemala. **Panama jaguar** (*P. o. centralis*), C America, Colombia. **Peruvian jaguar** (*P. o. peruviana*), Ecuador, Peru, Bolivia. **Amazon jaguar** (*P. o. onca*), forests of Orinoco and Amazon basins. **Paraná jaguar** (*P. o. palustris*), S Brazil, Argentina. **Arizona jaguar** (*P. o. arizonensis*), USA to NW Mexico. *P. o. veracrucensis* and *P. o. harnandes*, Mexico, very rare.

Snow leopard E
Panthera uncia

The Altai, Hindu Kush, and Himalayas. In mountain steppe and coniferous forest scrub at altitudes between 5,900 and 18,000ft. HBL 47–59in; TL about 35in. Coat: soft gray, shading to white on belly; head and lower limbs marked with

solid black or dark brown spots arranged in rows; body covered with medium brown blotches ringed with black or dark drown; a black streak along the back; tail round and heavily furred; ear pinnae black edged; winter coat lighter. Gestation: 98–103 days. Longevity: up to 15 years in captivity.

Clouded leopard V
Neofelis nebulosa

India, S China, Nepal, Burma, Indochina to Sumatra and Borneo, and Taiwan. In dense forest at altitudes up to 6,500ft. HBL 24–43in; TL 24–35in; shoulder height about 31in; wt 33–44lb. Coat: short, from dark brown or gray to brownish or ocher-yellow, patterned distinctively with black stripes, spots and blotches;

forehead and top of head lack spots, but six lines extend lengthwise across nape of neck, the outer ones much wider than the central ones; two stripes along the back which reduce to spots near the head; flanks with oblong to roundish blotches, each comprising a ring of pale fur outside a dark brown or grayish ring enclosing a paler center that is spotted; legs, throat and belly with black blotches shading to white on the underparts; ear pinnae rounded, black on outside and white on inside with a buff spot; tail long and bushy, ringed and tipped with black. Gestation: 85–90 days. Longevity: up to 17 years in captivity Subspecies: 2. **Formosan Clouded leopard** (*N. n. brachyurus*) from Taiwan, with tail not so long. All others included in *N. n. nebulosa*.

SMALL CATS

Genus *Felis* ✴
Twenty-eight species.
Family: Felidae.
Distribution: N and S America, Eurasia, Africa.

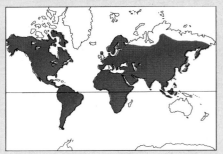

Habitat: from arid regions with sparse cover (Desert cat), through steppe, bush and savanna (African wild cat) to cool-temperate forest (European wild cat).

Size: head-body length from 14–16in (35–40cm) in the Black-footed cat to 41–77in (105–196cm) in the puma; weight from 2.2–4.4lb (1–2kg) to 227lb (103kg).

Coat: most often spotted or striped, sometimes uniform; face markings often striped with a black tear stripe from the eye.

Gestation: from about 56 days in the Leopard cat to 90–96 days in the puma.

Longevity: 12–15 years in the European wild cat.

Species include the **Wild cat** (*F. sylvestris*), W Europe to India (includes **European wild cat**), and subspecies **African wild cat** (*F. s. lybica*), Africa, and **Domestic cat** (*F. s. catus*), worldwide. **Leopard cat** (*F. bengalensis*), E and SE Asia. **Lynx** (*F. lynx*), Europe to Asia, N America. **Puma** (*F. concolor*) or **cougar**, Canada to Patagonia. **Bobcat** (*F. rufus*), Canada to Mexico. **Ocelot** [v] (*F. pardalis*), Arizona to Argentina. **Margay cat** [v] (*F. wiedi*), Mexico to Argentina. **Geoffroy's cat** (*F. geoffroyi*), S America.

✴ CITES listed. [v] Vulnerable.

▶ **A serval stalks through long grass.** This slender long-legged African savanna cat prefers to live near water. The small head with large rounded ears and the long neck are also characteristic.

▶ **Surprised by night,** a Scottish Wild cat puts on an impressive threat display.

ALTHOUGH most have never been properly studied, the 28 species of "small cats" are generally known to be similar in anatomy, morphology, biology and behavior to the better known "big cats."

Distinguishing characteristics of the genus *Felis* include: a fully ossified (bony) hyoid bone in the vocal apparatus which prevents them from roaring; claws that can be withdrawn (except in the Flat-headed cat) into sheaths which are longer on the outer side (of equal length in big cats); and a hairless strip along the front of the nose (furred in the big cats). When resting, they tuck their forepaws beneath their body by bending them at the wrist joint, and the tail is wrapped round their body; big cats at rest place their paws in front of their bodies and extend their tails straight behind them. Small cats feed in a crouched position whereas big cats lie down to feed.

The classification of small cats is controversial: here we have grouped the European wild cat, the African wild cat and the Domestic cat together as one species, *F. silvestris*. In the table overleaf the facts presented for *F. silvestris* are those for the European wild cat, which is taken to be representative of the group.

The Domestic cat (*F. s. catus*) is thought to be a descendant of the African wild cat (*F. s. lybica*), which was domesticated in Ancient Egypt, probably about 2000 BC. Controversy rages over the issue but, whatever its origins, the Domestic cat is certainly the most successful and widespread felid, being found worldwide in human settlements, often leading a wild (feral) existence.

There is little information to allow comparison of the behavior of the species and subspecies of small cats. However, in one recent study in Scotland, the social organization and feeding behavior of the European wild cat and the Domestic cat were compared. This revealed that the society of feral Domestic cats varied considerably: from solitary individuals to groups of up to 30 members, or a mixture of the two life-styles, depending on the availability of food and its dispersal in time and space. It was found that feral Domestic cats that depend on dispersed rabbit prey in an open habitat hunt alone, while those exploiting a clumped food resource, such as food put out at human dwellings, live in groups. The social organization of European wild cats was found to be very similar to solitary feral Domestic cats, although their ranges were on average larger at 435 acres (176 hectares) for an adult male compared to an average of 87 acres (35 hectares) for Domestic cats. This was due to more widely

Bobcat and Lynx: habitat and physique

Although very similar in basic form, different small cats species show a wide range of physical adaptations to their habitats. The lynx (1) and the bobcat (2) are two North American felids of similar size—11–66lb (5–30kg)—which occupy different habitats.

The plain brownish-gray coat of the lynx enables it to be inconspicuous against a background of dense, moss-laden coniferous forests and swamps—the typical vegetation from which it stalks its main prey, the Snowshoe hare. The black-spotted brown coat of the bobcat blends in well with the background of rocks, brush and other dense vegetation where its main prey—cottontails—feed. Because of the denser cover, sound may be more important than sight in locating prey for the lynx than the bobcat, and hence its ear tufts, which are thought to help hearing, are longer than those of the bobcat.

Lynx live in cold northern latitudes where snow lies deep for much of the year. As adaptations to the lower temperatures (to −70°F/−57°C) they have shorter tails than the bobcats and their foot pads are well protected with a dense covering of fur, while those of the bobcat are bare. The longer legs of the lynx are also an adaptation to traveling through deep snow, where the bobcat is at a disadvantage. **TNB**

dispersed food resources; for example, rabbit prey were more sparsely distributed in the high altitude, young forest and scrub of scottish wild cats' habitat than in the farmland habitat of the feral domestic cats. The basic hunting technique in dense habitats with defended territories where prey was scarce, small and widely dispersed, was solitary stalking. However, where prey was abundant, relatively large and patchily distributed, cats often lived in groups to defend and exploit this food. This detailed study of the wild and domestic subspecies of *F. silvestris* exemplifies the functional relationship between grades of social organization and feeding ecology, where both abundance and dispersion of food are important.

The most serious threat to small cats is the fur trade, which continues to demand large numbers of spotted cat skins, despite considerable adverse public opinion. The resulting pressure on wild populations of these rare, beautiful and little understood creatures is pushing many of them to the brink of extinction. To take an example, the ocelot is particularly vulnerable, its skin being in great demand by the fur industry. Widespread, and also easily trapped or shot, it is consequently the most frequently hunted small cat in Latin America. In 1975, Britain alone imported 76,838 ocelot skins. It is now rare and threatened in parts of its range, and populations everywhere are seriously reduced in number.

The bobcat, lynx and puma are suffering similar fates, and the situation is often exacerbated by their persecution as pests and by destruction of their habitat. In North America alone, during 1977–78 over 85,000 bobcat skins and 20,000 lynx skins were harvested, together worth over $16 million.

The situation with many of the smaller cats is even worse: for example, over 20,000 pelts of the rare Geoffroy's cat are taken each year. Similarly, both the Margay cat and Leopard cat have beautifully marked coats that are much in demand by the fur trade, and both are in consequence subject to intense hunting.

Enormous numbers of felid skins are required because of the intricate matching procedure required for each garment. When a species becomes too scarce to provide the minimum number of skins demanded by the trade, another more common one is exploited by the illegal hunters. Thus, species by species, the small spotted cats are being hunted to a point where the remaining populations are so small and widely dispersed that they may never recover. **GK**

THE 28 SPECIES OF SMALL CATS

Abbreviations: HBL = head-body length; TL = tail length; wt = weight.

[V] Vulnerable. [R] Rare. [E] Endangered. [I] Threatened, but exact status indeterminate.

Genus *Felis*

Mostly forest-dwellers preying on small mammals, but opportunistically taking any small vertebrate prey. Coat variable in density and length, most often spotted or striped, but sometimes uniform. Face markings variable but often striped; black "tear" stripes normally present. Tail tapered or rounded at tip, often with dark rings. Ear pinnae vary in size and degree of roundness, with tufted tips in some species. Color of iris of eyes varies from rich orange, through yellow to green; pupils contract to a circle, slit or spindle shape. Skull normally rounded. The anatomy, morphology and biology of wild members of the genus is poorly understood. (The following table cites features where they are known.)

African golden cat
F. aurata

Senegal to Zaire and Kenya. Forest and dense scrubland. Prey: small mammals and birds. HBL 27–37in; TL 11–14in; wt 30–40lb. Coat: chestnut-brown to silver-gray; patterning variable in type and extent; eyes brown; tail tapered at tip; ears small and rounded.

Asiatic golden cat [I]
F. temmincki
Asiatic golden or Temminck's golden cat.

Nepal to S China and Sumatra. Forest. Prey: rodents, small deer, game birds. HBL 29–41in; TL 16–22in; wt 13–24lb. Coat: uniform golden-brown with head typically striped with white, blue and gray (much variation). Litter 2–3.

(**Black-footed cat** continued)
on underside of feet; legs with wide black rings on upper parts; skull broad; ears large; pupils contract to slit; hair on soles of feet. Gestation 63–68 days; litter 2–3.

Bobcat
F. rufus
Bobcat or Red lynx

S Canada to S Mexico. Rocky scree, rough ground, thickets, swamp; Prey: rodents, small mammals, large ground birds; active at twilight. HBL 24–42in; TL 4–8in; wt 13–68lb. Coat: barred and spotted with black on reddish-brown (very variable) basic color: underside white, tail tip black; heavily built with a short tail and short ear tufts. Gestation 60–63 days; litter 1–4.

Caracal
F. caracal
Caracal, lynx or African lynx.

Africa and Asia from Turkestan, NW India to Arabia. Wide habitat tolerance. Prey: rodents and other small mammals including young deer, which are either run down or pounced on; mainly active at twilight but will hunt during the night in hot weather or by day in the winter. HBL 22–29in; TL 9in; wt 35–51lb. Coat: reddish-brown to yellow-gray; underside white; ears tufted; legs very long; eyes yellow-brown; pupils contract to a circle. Gestation 70–78 days; litter 1–4.

Chinese desert cat
F. bieti

C Asia, W China, S Mongolia. Steppe and mountain. HBL 27–33in; TL 12–14in; wt about 12lb. Coat: brownish-yellow with dark spots merging into stripes; underside paler;

Flat-headed cat [I]
F. planiceps

Borneo, Sumatra, Malaya. Forest and scrub; prefers proximity to water. Prey: small mammals, birds, fish, amphibians; nocturnal. HBL 16–20in; TL 5–6in; wt 12–18lb. Coat: plain reddish-brown, underside white; dark spots on throat, belly, inner sides of legs; ears black with ocher spot at the base; tear streaks white; head slightly flattened; legs short; paws small; ears small and rounded; claws not fully retractile.

Geoffroy's cat
F. geoffroyi
Geoffroy's cat/Geoffroy's ocelot.

Bolivia to Patagonia. Upland forests and scrub. Prey: birds and small mammals; climbs and swims well. HBL 18–27in; TL 10–14in; wt 4–8lb. Coat: silver-gray, through ocher-yellow to brownish-yellow with small black spots. Litter 2–3.

Jaguarundi [I]
F. yagouaroundi
Jaguarundi, jaguarondi, eyra, Otter-cat.

Arizona to N Argentina. Forest, savanna, scrub. Prey: birds, rabbits, rodents, frogs, fish, poultry; active at twilight. HBL 21–26in; TL 13–24in; wt 12–22lb. Coat: either uniform red or uniform gray, lighter underneath; newborn dark spotted; legs very short; body long and slender; ears small, round; eyes brown; pupil contracts to a slit. Gestation 63–70 days; litter 2–4.

Iriomote cat [E]
F. iriomotensis

Iriomote Islands, Ryukyu Islands, Sub-tropical rain forest, always near water. Prey: waterbirds, small

▲ **A stalk, pounce and kill sequence** showing 12 species of small cat, arranged in a west (America) to east (Asia) order reflecting distribution.
(1) Ocelot. (2) Margay cat.
(3) Tiger cat. (4) Jaguarundi.
(5) and (6) European and African Wild cat. (7) Black-footed cat.
(8) Sand cat. (9) Jungle cat.
(10) Leopard cat. (11) Asiatic golden cat. (12) Fishing cat.

Bay cat [R]
F. badia
Bay or Bornean red cat.

Borneo. Rocky scrub: Prey: small mammals and birds. HBL about 20in; TL about 12in; wt 4–7lb. Coat: uniform bright reddish-brown; lighter colored on underside; head short and rounded.

Black-footed cat
F. nigripes

S Africa, Botswana, Namibia. Steppe and savanna. Prey: rodents, lizards, insects. HBL 14–16in; TL 6–7in; wt 2–4lb. Coat: light brown with dark spots on body and black patches

red tinge on the back; tail ringed; skull broad; ears large; soles of the feet padded with fur.

Fishing cat
F. viverrina

Sumatra, Java, to S China to India. Forest, swamps, marshy areas (dependent on water). Prey: fish, small mammals, birds, insects and crustacea. HBL 22–33in; TL 8–12in; wt 12–18lb. Coat: short and coarse, light brown with dark brown or black spots; tail ringed with black; paws slightly webbed and claws not fully retractile. Gestation 63 days; litter 1–4.

rodents, crabs, mud-skippers; nocturnal and strictly territorial with ranges up to 0.75 sq mile.

Jungle cat
F. chaus

Egypt to Indochina and Sri Lanka. Dry forest, woodland, scrub, reed beds, often near human settlements. Prey: rodents and frogs, occasionally birds; most active in day. HBL 24–30in; TL 10–14in; wt 15–30lb. Coat: sandy-brown to yellow-gray, sometimes with dark stripes on face and legs and with a ringed tail; young have distinct close-set striped pattern which disappears in the adult, tail short; legs

(Jungle cat continued)
long; ears tapered and tufted with a light spot at the base. Gestation 66 days; litter 2–5.

Kodkod
F. guigna
Kodkod, huiña.

C and S Chile, W Argentina. Forest. Prey, birds and small mammals; probably nocturnal. HBL 16–20in; TL 7–9in; wt 4–7lb. Coat: gray varying to ocher-brown, with dark spots and ringed tail; underside whitish; prominent dark band across the throat, but few markings on the face.

Leopard cat
F. bengalensis
Leopard or Bengal cat.

Sumatra, Java, Borneo, Philippines, Taiwan, Japan. Forest, scrubland, particularly near water. Prey: rodents, small mammals, birds, which it drops on from above; active at night and at twilight; good swimmer and climber. HBL 14–24in; TL 6–16in; wt 8lb. Coat: background color varies from ocher-yellow to ocher-brown with underside paler; covered with black spots; prominent white spot between the eyes; eyes yellow-brown to greenish-yellow; ears rounded and black with a white spot on the outer surface. Gestation about 56 days; litter 2–4.

Lynx
F. lynx (F. pardina)
Lynx, Northern lynx.

W Europe to Siberia; Spain and Portugal; Alaska, Canada, N USA. Coniferous forest and thick scrub. Prey: rodents, small ungulates;

crepuscular. HBL 26–43in; TL 2–7in; wt 11–64lb. Coat: light brown with dark spots; tail black-tipped (coloration and patterning very variable); ear tufts long and black; two tassels on throat; tail short; paws large with thick fur padding; pupils contract to a circle; 28 teeth. Gestation 60–74 days; litter 1–5.

Marbled cat
F. marmorata

Sumatra, Borneo, Malaya to Nepal. Forest. Prey: rodents, birds, small mammals, insects, lizards, snakes; nocturnal and arboreal. HBL 16–24in; TL 18–21in; wt about 12lb. Coat: soft, long fur, light-

(Marbled cat continued)
brown with striking patterns of dark brown blotches and spots all over.

Margay cat [V]
F. wiedi
Margay cat, "tigrillo".

N Mexico to N Argentina. Forest, scrubland. Prey: rats, squirrels, opossums, monkeys, birds; excellent climber. HBL 18–27in; TL 14–20in; wt 9–20lb. Coat: yellow-brown with black spots and stripes; tail ringed; eyes large, dark brown. Litter 1–2.

Mountain cat [R]
F. jacobita
Mountain or Andean cat.

S Peru to N Chile. Mountain steppe. Prey: small mammals and birds. HBL 27–29in; TL about 18in; wt 8–15lb. Coat: brown-gray with dark spots, a ringed tail and white belly; long- and thick-haired especially on the tail, which appears perfectly round.

Ocelot [V]
F. pardalis

Arizona to N Argentina. Forest and steppe. Prey: small mammals, birds, reptiles; excellent climber and swimmer; may live in pairs. HBL 25–38in; TL 11–16in; wt 24–35lb. Coat: ocher-yellow to orange-yellow in forested areas, grayer in arid scrubland; black striped and spotted; underside white; tail ringed; eyes brownish; hair curls at the withers to lie forward on upper neck. Gestation 70 days; litter 2–4.

Pallas's cat
F. manul
Pallas's cat, manul.

Iran to W China. Mountain steppe, rocky terrain, woodland. Prey: mainly rodents. HBL 20–25in; TL 8–12in; wt 7–11lb. Coat: long, orange-gray with black and white head markings; belly light gray; ears small and rounded, widely separated on a broad head with a low forehead; pupil contracts to a circle; front premolar teeth missing, giving 28 teeth; eyes face almost directly forward. Litter 1–5.

Pampas cat
F. colocolo

Ecuador to Patagonia. Grassland, forest, scrub. Prey: small- to medium-sized rodents, birds, lizards large insects; probably nocturnal. HBL 20–27in; TL 11–13in; wt 8–14lb. Coat: long, soft, gray-brown with brown spots (very variable) and with reddish-hue; ears tapered and tufted; eyes yellow-brown; pupil contracts to a spindle. (Previously called *F. pajeros*, derived from the Spanish "paja," meaning straw, because it lives in reed beds.)

Puma
F. concolor
Puma, cougar, Mountain lion, panther.

Includes Eastern cougar [E] (*F. c. cougar*), E N America, and Florida cougar [E] (*F. c. coryi*), S Canada to Patagonia. Forest to steppe, including conifer, deciduous and tropical forests, grassland and desert. Prey: from small rodents to fully grown deer; mainly active at twilight. HBL 41–77in; TL 26–31in; wt 79–227lb. Coat: plain gray-brown to black (very variable); cubs initially dark spotted; head round and small; body very slender; eyes brown; pupils circular; tail black-tipped. Gestation 90–96 days; litter 3–4.

Rusty-spotted cat
F. rubiginosus

S India and Sri Lanka. Scrub, forest, around waterways and human settlements. Prey: small mammals, birds, insects. HBL 14–19in; TL 6–10in; wt 2–4lb. Coat: rust colored with brown blotches and stripes.

Sand cat
F. margarita

N Africa and SW Asia (Sahara to Baluchistan). Desert. Prey: small rodents, lizards, insects; nocturnal. HBL 16–22in; TL 10–14in; wt 4–5.5lb. Coat: plain yellow-brown to gray-brown; tail ringed, with a black tip; kittens born with distinct coat markings which usually fade in adulthood; hair covers paw pads; head very broad; eyes large and forward on the head; ears tapered.

Serval
F. serval

Africa. Savanna, normally near water. Prey: game birds, rodents, small ungulates; good climber. HBL 27–39in; TL 14–16in; wt 30–42lb. Coat: orange-brown with black spots (very variable); slender build with long legs, small head, rather long neck and large, rounded ears; eyes yellowish; pupils contract to a spindle. Gestation about 75 days; litter 1–3.

Tiger cat [V]
F. tigrinus
Tiger or Little spotted or Ocelot cat, oricilla.

Costa Rica to N Argentina. Forest. Prey: small mammals, birds, lizards, large insects; good climber. HBL 16–22in; TL 10–16in; wt 4–8lb. Coat: light brown with very dark brown stripes and blotches; underparts lighter; white line above the eyes. Gestation 74 days; litter 1–2.

Wild cat
F. silvestris
(Includes Domestic cat, *F. s. catus*, and African wild cat, *F. s. lybica*)

W Europe to India; Africa (*F. s. catus* worldwide—introduced by man). Open forest, savanna, steppe. Prey: small mammals and birds; nocturnal. HBL 20–31in; TL 11–14in; wt 7–13lb (slightly smaller for *F. s. lybica*). Coat: medium brown, black-striped; *F. s. lybica* is light brown with stripes; *F. s. catus* shows many color forms; females generally paler than males; tail black-tipped. Gestation 68 days; litter 3–6.

GK

North America's Secretive Cats

Flexible land use in the lynx, bobcat and puma

Three main species of small cats inhabit the wilderness areas of North America. Distribution of the lynx, in the coniferous forest and thick scrub of Alaska, Canada and the northern USA, barely overlaps with that of North America's most common felid, the bobcat, which extends south across the USA including most habitats (except those without sign of tree or shrub). The much larger puma, cougar or mountain lion inhabits mostly rocky terrain ranging from forest to desert from southern Canada south to Patagonia in southern South America.

Adapting to a wide range of habitats, climate and degrees of prey availability, the solitary individuals of each species display great flexibility of behavior. (See also the physical adaptations described on p43.) Their success in breeding and survival is based on land tenure—the maintenance of more or less exclusive access to prey within a defined home area. Generally, older resident cats which occupy distinct areas year after year have the greatest breeding success. Others, usually younger individuals, are not as successful because they fail to occupy areas in prime habitat permanently, unless a vacancy occurs or they settle in less favorable habitats.

Of the many influences on the land-tenure system of cats, abundance, population stability and distribution and mobility of prey are especially important. Abundant prey allows cats to survive on smaller areas at higher densities. Stable prey populations permit long-term familiarity with hunting areas and neighbors, so prompting stable tenure; scattered, fluctuating prey may force overlap and the sharing of limited resources. Finally, highly mobile prey may demand seasonal shifts of hunting areas. To adapt to all these factors requires a flexible land-tenure system.

The puma, largest of all the "small cats," demonstrates this flexibility. In the rugged mountains of northwestern North America, individual pumas shift between large summer ranges (which vary from 41–80sqmi/ 106–207sqkm for females and up to 113sqmi/293sqkm for males) and smaller winter ranges averaging 42sqmi (107sqkm) and 49sqmi (126sqkm) for females and males, respectively. Locations of seasonal ranges are determined by the immigration patterns of their main prey, Mule deer and elk. Winter snow at higher altitudes forces hoofed mammals, and consequently the pumas, from higher summer ranges into lower valleys, but pumas maintain approximately the same distance from each other in both their summer and winter ranges. The home ranges of adult male pumas overlap a little, those of adult females overlap more, sometimes completely, and one male's range may overlap those of several females. Sometimes a female's

▲ **Puma surveying its territory.** The puma is the largest of North American cats and it may travel through summer ranges of over 77–116sqmi (200–300sqkm).

Warning Off Intruders

North American cats advertise land occupancy largely by scent and visual signals. The means used to warn off intruders include the depositing of urine, feces, and anal gland secretions, and making scrapes in the ground. Vocalizations play little part.

Bobcats frequently squirt urine (1) along common travel routes (which in females may also indicate whether they are receptive or approaching receptivity), deposit feces (2) in latrine sites (middens) which if near dens may indicate that they are being used by females with offspring, and (3) scrape with or without defecating or urinating along trails, trail intersections and other important places in their home ranges; scraping also adds to the conspicuousness of other scent marks. Intruding cats coming across such marks (4) usually respect the prior rights of other cats, as newcomers seldom succeed in permanently settling in areas already occupied by residents. Body posture and facial expressions are probably effective close-range signals when two animals meet, as in (5) a defensive threat, and (6) an attack threat.

Some of the bobcats travelled at least 20mi (30km) away in search of more prey. Young bobcats dispersed up to 99mi (158km) from their place of birth. Adult bobcats, like adult pumas, avoid encounters except when mating.

Less is known about the land-tenure system of the lynx. The availability of their main prey, Snowshoe hares, appears to determine the spatial organization of lynx. Where hares are abundant, lynx may have a system similar to the bobcat's, with males using ranges up to 19sqmi (50sqkm) and females using ranges up to 10sqmi (26sqkm). However, when hare populations crash, a periodic occurrence in the north, home ranges enlarge and overlapping increases as lynx seek out remaining concentrations of hares. When hares suddenly decline over a vast area, lynx have been known to disperse up to 300mi (480km) and areas of use become large, up to 47sqmi (122sqkm) and 94sqmi (243sqkm) for females and males, respectively. Long-term stable land tenure is unlikely to occur among fluctuating lynx populations because long-lived neighboring residents familiar with each other and stable prey populations are needed, and these conditions occur only periodically in many harsh northern environments.

Land tenure in the ocelot and jaguarundi, which occur in the southern USA at the northern limit of their ranges, has not been intensively studied, but one early account suggests that an ocelot territory never contains more than one male and one female. Because their tropical environment is more stable than northern environments, and because they feed on a wide variety of prey (mice, rats, pacas, agoutis, coatis, monkeys and peccaries), ocelots should have relatively stable home ranges.

Systems of land tenure serve many functions. Some maintain densities of breeding adults below levels set by food supplies and thus regulate populations. Where deer and elk increased from 1.6–2.1 and from 0.9–1.5 per sqmi respectively, puma densities remained constant at 1 puma per 14sqmi (35sqkm) throughout a five-year period. Land-tenure systems also ensure reproductive success by maximizing the number of females bred by individual males. In one area, there were 1.7 resident females for each resident male bobcat. Survival is probably enhanced because residents familiar with locations of prey and cover have lower death rates and produce more young than individuals who are unable to occupy land permanently. TNB

▼ **Homes ranges of the bobcat.** The size of home ranges in bobcats varies seasonally with prey availability. In this area, following a decline in the local rabbit population, one female bobcat extended her range. Overall, male ranges overlap very little with those of other males (one overlap area shown shaded), but partially embrace the ranges of several females. Dens in active use are marked with fecal middens (heaps).

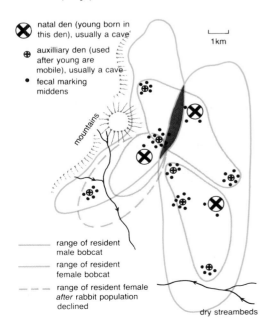

range may be partially overlapped by more than one male, but which male mates with the female is not known. Where ranges overlap, pumas seldom use the same localities at the same time. Avoidance of other pumas, a frequent but little understood behavior pattern, also reduces fighting between these powerful predators to rare events. Young pumas disperse up to at least 28mi (45km) from their natal areas.

The land-tenure system of bobcats appears to be highly flexible, varying with different habitats and prey—mainly hares or rabbits. In one area where rabbits were relatively abundant but protective cover limited, adult male bobcats occupied partially overlapping ranges up to 42sqmi (108sqkm), while adult females occupied smaller, more exclusive areas up to 17sqmi (45sqkm). A natal den and up to five auxiliary dens which were used to rear young formed focal points of activity within each female's range. Less than one percent of the average area used by adult females was shared with other resident females and about two percent of the area used by males was shared with other resident males.

Like those of male pumas, the home ranges of resident male bobcats overlap those of one or more resident females. This permits them to mate with as many females as possible, while keeping other males away. If rabbit populations remain stable, bobcats roam the same ranges year after year, but if rabbits decline, resident bobcats have to enlarge or abandon their familiar ranges.

THE DOG FAMILY

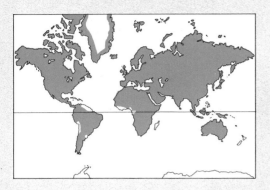

Family: Canidae
Thirty-five species in 10 genera.
Distribution: worldwide, excluding a few areas (eg Madagascar, New Zealand) in most of which Domestic dog introduced.

Habitat: evolved in open grasslands but now adapted to an exceptionally wide range of habitats.

Size: ranges from the Fennec fox—minimum adult head-body length 9.5in (24cm), tail 7in (18cm) and weight 1lb (0.8kg)—to the Gray wolf, up to 6ft 7in (200cm) in overall length and 175lb (80kg) in weight. Other fox species weigh between 3.3 and 20lb (1.5–9kg), all other species between 11 and 60lb (5–27kg).

Gray wolf *Canis lupus*
Red wolf *Canis rufus*
Coyote *Canis latrans*
Dingo *Canis dingo*
Domestic dog *Canis familiaris*
Jackals Four species of *Canis*

Vulpine foxes Twelve species of *Vulpes*
South American foxes Seven species of *Dusicyon*
Arctic fox *Alopex lagopus*
Bat-eared fox *Otocyon megalotis*

African wild dog *Lycaon pictus*
Dhole *Cuon alpinus*
Maned wolf *Chrysocyon brachyurus*
Raccoon dog *Nyctereutes procyonoides*
Bush dog *Speothos venaticus*

CANIDS evolved for fast pursuit of prey in open grasslands and their anatomy is clearly adapted to this life. Although the 35 species and 10 genera vary in size from the tiny Fennec fox to the powerful Gray wolf, all but one have lithe builds, long bushy tails, long legs and digitigrade four-toed feet, tipped with non-retractile claws. The Bush dog is the single exception. A vestigial first toe (pollex) is found on the front feet in all save the African wild dog. The dingo and the Domestic dog also have vestigial first claws (dew claws) on their hind legs. Other adaptations to running include fusion of wrist bones (scaphoid and lunar) and locking of the front leg bones (radius and ulna) to prevent rotation.

Male canids have a well-developed penis bone (baculum) and a copulatory tie keeps mating pairs locked together, facing in opposite directions for a time lasting from a few minutes to an hour or more. Its function is unknown, although people speculate loosely that it serves to "cement the pair bond." The mechanism involves a complex arrangement of blood vessels and baculum, combined with the reversed body position, so that blood is trapped in the engorged penis, impeding withdrawal.

Gestation lasts about nine weeks in most species and one litter is born annually. The pups' eyes generally open at about two weeks and the young begin to take solid food from 2–6 weeks of age. Solid food is regurgitated by *Canis* species, African wild dogs, dholes and probably Maned wolves but carried to cubs by vulpine and Arctic foxes.

Canids originated in North America during the Eocene (54–38 million years ago), from which five fossil genera are known. Two forms, *Hesperocyon* of North America and *Cynodictis* of Europe, are ancient canids, with civet-like frames. They share this long-bodied, short-limbed physique with the

Miacoidea from which all Carnivora evolved (see p14). As modern canid features evolved, the family blossomed: 19 genera in the Oligocene (38–26 million years ago), 42 in the Miocene (26–7 million years ago), declining to the 10 genera recognized today.

The heel of the carnassial teeth in most canids has two cusps, but in the Bush dog, African wild dog and dhole only one, and this has led some taxonomists to group these genera together as the subfamily Simocyoninae, distinct from the subfamily Caninae, containing all other species except the Bat-eared fox, classified alone in the Otocyoninae. Members of the three largest genera, *Canis*, *Vulpes* and *Dusicyon*, are generally more similar to members of their own genus than to members of others, but the distinctions between genera are often minimal. In descending order of atypicality

▼ **Members of the genus *Canis*,** whose breeding plasticity (especially in the Gray wolf, *Canis lupus*) is the source of all today's breeds of Domestic dog (*C. familiaris*). (**1**) Coyote (*C. latrans*) showing "play" face; 35lb (16kg). (**2**) Red wolf (*C. rufus*) in submissive greeting posture 50lb (23kg). Subspecies of Gray wolf: (**3**) Arabian wolf in defensive threat posture; 50lb (23kg). (**4**) Mexican wolf in offensive threat posture; 70lb (32kg). (**5**) European wolf howling; 55lb (25kg). (**6**) Tibetan wolf cocking leg (therefore, a dominant individual) to urinate. (**7**) Gray wolf/Husky cross, a common wolf/Domestic dog hybrid; 120lb (54kg).

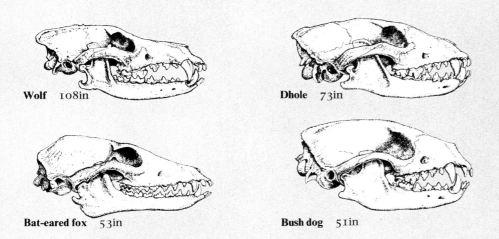

Wolf 108in

Dhole 73in

Bat-eared fox 53in

Bush dog 51in

Skulls of Canids

Canids have long muzzles, well-developed jaws and a characteristic dental formula (see p12) of I3/3, C1/1, P4/4, M2/3 = 42, as exemplified by the Gray wolf. Three species depart from this pattern, the Bat-eared fox (= 48), the dhole (= 40), and the Bush dog (= 38). Shearing carnassial teeth (P4/M1) and crushing molars are well developed and, except in the Bat-eared fox, are the largest teeth.

5

6

7

the most unorthodox canids are the African wild dog, Bush dog, Bat-eared fox, Raccoon dog, dhole, the Maned wolf and the Arctic fox. These seven are placed in single-species genera, but as none is any more closely related to any other than to the bulk of canids, it seems prudent to abandon the division into subfamilies.

The most striking feature of the canids is their opportunistic and adaptable behavior.

This is most conspicuous in the flexible complexity of their social organization. Remarkably, in this respect there is as much variation within as there is between species. Although African wild dogs, and possibly dholes and Bush dogs, almost always hunt in packs, wolves, coyotes and jackals feed on prey ranging from ungulates to berries. Partly as a consequence, they lead social lives that vary from solitary to sociable—some wolf populations comprise monogamous pairs, whilst elsewhere packs may have up to 20 members. These species, and some others like Red and Arctic foxes, live in groups even where large prey does not abound and where they hunt alone. Apart from hunting, there are many other reasons for group living—cooperative defense of territories or large carcasses, communal care of offspring ("helping," as in jackals, see p58), rivalry with neighboring groups.

Selective breeding has emphasized canid plasticity in creating the Domestic dog. Various origins have been proposed for Domestic dogs and doubtless many canids have been partly domesticated at one time or another, but the wolf is generally accepted as ancestor of today's Domestic dogs. The earliest Domestic dog is either the so-called Starr Carr dog of Yorkshire, England, which lived 9,500 years ago, or, perhaps, the specimen found by Coon in Iran, dating back over 11,000 years.

With all their adaptability, members of the dog family cannot escape the indirect threat of habitat destruction. The Small-eared dog and the Bush dog are seen so rarely that there is very little information on either. The Simien jackal numbers less than 500 individuals in the highlands of Ethiopia, and the Maned wolf only 1,000–2,000 in its Argentinian and Brazilian strongholds. These species, and perhaps the African wild dog, are probably endangered. DWM

WOLVES

Two of 9 species of the genus *Canis*
Family: Canidae.
Distribution: N America, Europe, Asia.

Gray wolf V

Canis lupus
Gray wolf, wolf, Timber or White wolf.

Distribution: N America, Europe, Asia, Middle East.

Habitat: forests, taiga, tundra, deserts; plains and mountains.

Size: head-body length 40–58in (100–150cm); tail length 13–20in (31–51cm); shoulder height 26–32in (66–81cm); weight 27–175lb (12–80kg). Males larger than females.

Coat: usually gray to tawny-buff, but varies from white (in N tundra) through red and brown to black; underside pale.

Gestation: 61–63 days.

Longevity: 8–16 years (to 20 in captivity).

Subspecies: up to 32 have been described. Surviving subspecies include the **Common wolf** (*C. l. lupus*), forests of Europe and Asia, medium-sized with short, dark fur. The **Steppe wolf** (*C. l. campestris*), steppes and deserts of C Asia, small, with a short, coarse gray-ocher coat. The **Tundra wolves** of Eurasia (*C. l. albus*) and America (*C. l. tundarum*), large, with coat long and light-colored. **Eastern timber wolf** (*C. l. lycaon*), once the most widespread in N America, but now only in areas of low human population density; smaller, usually gray in color. The **Great Plains wolf** EX or Buffalo wolf (*C. l. nubilus*), white to black in color, once followed the great herds of bison on the North American plains.

Red wolf E

Canis rufus

Distribution: SE USA (now probably extinct in wild).

Habitat: coastal plains, forests.

Size and weight —33–66lb (15–30kg)— intermediate between Gray wolf and coyote (pp48–49); gestation and longevity same as Gray wolf.

Coat: cinnamon or tawny with gray and black highlights.

V Vulnerable. E Endangered. EX Extinct.

No animal is more enshrined in the myths and legends of the peoples of the North than the wolf. For thousands of years it has competed with man for game and killed his farm animals. Stories of attacks on humans are rife, but very many are exaggerated, and most are fantasy. There are few substantiated cases of a healthy wolf attacking a man and even today a sturdy stick is all that shepherds in Italy's Abruzzi mountains require to fend off a threatening wolf. Many legends probably originated at times of war, famine or epidemic, when wolves scavenged corpses. Wolves can, however, wreak havoc on farm stock—there are still occasions in Italy when over 200 sheep may be killed by a pack in one night.

Two species of wolf remain today. The Gray wolf is the largest member of the dog family. It was once the most widespread mammal, apart from man, outside the tropics (the Red fox has that distinction today). Now it is restricted to a few large forests in eastern Europe, some isolated mountain refuges in the Mediterranean region, mountains and semidesert areas of the Middle East, and wilderness areas throughout Asia and North America. This decline appears to be largely the result of human persecution and habitat destruction. The extremely rare Red wolf was once found throughout the southeastern USA but is now thought to be extinct in the wild, the result largely of hybridization with eastward-moving coyotes (see p54).

The immense former geographic range and ecological adaptability of the wolf are reflected in wide variations of form and behavior. The smallest and lightest-colored wolves inhabit semidesert areas. Forest-dwelling wolves tend to be of medium size and grayish. Tundra wolves are among the largest, their coat color ranging from white through gray to black.

Wolves will take a wide range of food. In wilderness areas, typical prey are moose, deer and caribou, which weigh up to 10 times as much as a wolf and are hunted by

An Annual Cycle of Stability and Aggression

Two wolves meet nose to nose, others lie sleeping, or observing pups at play. The wolf pack appears calm, friendly, harmonious. Only the informed observer (and wolves!) will notice the signs indicating the hierarchy and tension: a tail kept somewhat lower and ears held back as a subordinate wolf greets a more dominant pack member (**1**), the non-aggressive play face of one pup as it is grabbed by another (**2**), a stiff-legged shove as one pup tries to exert authority over one of its contemporaries (**3**).

From late fall to late winter, when breeding starts, such interactions become more frequent and purposive. Only one female usually mates and females may fight vigorously for that right. Even if the dominant female is not challenged, fights may break out among low-ranking females seeking a higher place in the hierarchy. Juveniles and pups may join fights and some pack members may be driven out. Males may also fight to become the pack sire. Eventually, the dominant pair copulate, generally two or three times a day for approximately 14 days. After that, calm is restored. With the hierarchy fixed for several months, some subordinate pack members may decide to leave.

packs. Young and old or otherwise debilitated animals are most likely to be taken. Smaller mammals such as beaver and hares may be important prey, for example in summer. When domestic animals are available, wolves often turn to them because they have poor defenses. Wolves will on occasion take carrion and plant material, and even when wild prey is available will scavenge at trash cans and refuse dumps.

To find enough food, wolf packs require extensive home ranges, from 40sqmi or less to areas in excess of 400sqmi (100–1,000sqkm), depending mainly on prey density. Scent marking and vocalizations (the long, deep "mournful" howls that together with yips, barks, growls and whines make up the wolf vocabulary) help to define and defend these territories (see pp52–53). Most packs in forested areas occupy stable, year-round territories. In northern tundra regions, packs are usually nomadic as they follow the migrations of caribou and saiga antelope, but even they may return each year to established summer denning areas.

The nucleus of the wolf pack is the breeding pair (wolves usually mate for life). Body postures are an important part of the "language" that creates and reinforces hierarchy in the pack. Pack size depends on the size and availability of prey. In moose country, packs of up to 20 individuals occur, but this declines to about 7 where the main prey are deer. In the Italian Abruzzi, where most natural prey has been exterminated, wolves subsist on human garbage, supplemented by the odd sheep, goat or dog. There, wolves mostly travel alone.

Breeding occurs in late winter and some 4–7 blind and helpless pups are born in a den. After about a month, the pups emerge, to receive food and attention from their parents and other pack members. If the food supply is ample, these "helpers" can assist the pups to be fit and large enough to travel with the pack at 3–5 months; if food is scarce, their presence may reduce the pups' chances of survival. Some pups leave the pack during the following breeding season, while others remain as "helpers." Wolves are sexually mature at two years old.

It is a paradox that the wolf, once so feared and hated, now finds its own destiny in man's hands. It is ironical, too, that today the wolf should be so outnumbered by its descendant, the domestic dog. The battle for survival has been won by man and only if man provides suitable refuges and accepts a small loss of livestock will the wolf survive.

▲ **Lone Gray wolf on the move.** Most wolves live in packs. Lone wolves are generally younger animals in search of their own territory and mate. The lone wolf skirts the territories of others and rarely howls or scent marks.

▶ **Hierarchy within a wolf pack** is well defined among both males and females. Changes in leadership result in social turmoil as pack members gang up on deposed dominants. At other times a gaping threat, as illustrated, suffices to maintain everyday relationships.

To Howl or Not to Howl?

How wolves keep their distance

On occasion, wolf packs meet. When they do, a fight may develop, with the common outcome a dead wolf, lying where it fell victim to the other pack, a snarl on its face as testimony to its violent death.

Such encounters, although rare, have over the past decade been the primary natural cause of death among adult wolves in the Superior National Forest of Minnesota, USA. The likelihood of such disastrous encounters is reduced by each pack restricting its movements to a relatively exclusive territory of 25–115sqmi (65–300sqkm). The territory may be 6–12mi (10–20km) across, but only the outer kilometer or so is shared with neighboring packs or lone wolves. This periphery is visited much less often than the rest of the territory, presumably because of the danger of accidentally running into hostile neighbors. How do wolves recognize this periphery and thus avoid it and their neighbors?

Scent marking provides part of the answer. The dominant animals of each pack urinate on objects or at conspicuous locations about once every three minutes as the pack travels about its territory. The density of scent marks in the border regions is twice as high as elsewhere, the reason for which is not fully understood. However, wolves are known to increase their rate of scent marking after they encounter scent marks left by strangers, as they do much more frequently on the occasional visits to the edges of their pack territory. This higher density of scent marks, both its own and those of strangers, appears to enable a pack to recognize the periphery and keep from trespassing into even more dangerous areas beyond.

However, scent marks only inform a pack of where its neighbors *were*, and approximately *when* they passed through, not about where they are now. If the neighboring pack is traveling in the border area along the same trail but in the opposite direction, then whatever the density of scent marks they alone will not prevent an accidental meeting. Wolf packs need a more instantaneous spacing mechanism, and howling appears to fill this requirement.

When a pack howls, all members usually join the chorus. Under ideal conditions, this chorus can advertise a pack's location over distances as great as 10km (6mi). Thus a pack can broadcast its whereabouts over much of its territory in an instant. When two packs are approaching one another along a common border the chances of hearing howling continue to improve: the more likely an encounter becomes, the better the chance of hearing the neighbor pack's howling, and when they do so wolves normally avoid the meeting.

We might expect a pack to howl frequently as it travels about its territory and to reply immediately upon hearing strangers howl nearby. In fact, howling occurs only sporadically (in one study, only once every 10 hours), and as often as not a pack declines to answer a stranger's howling. The reasons for this apparent reluctance lie in another type of encounter between packs. On several occasions, packs in Superior National Forest have been observed to invade a neighbor's territory, follow a trail straight to the residents' location, and then attack them. In at least two of these "deliberate" encounters, resident wolves died and, in one case, the resident pack apparently disbanded after the attack and its territory was usurped by the intruders. Thus, a reply may incite an attack.

Although these deliberate encounters are

▲ **In single file,** a pack of wolves sets off in search of prey. Their travels may span up to (400sq mi) 1,000sq km as they hunt cooperatively for prey as large as moose.

▶ **Howling wolves** advertise their presence in order to avoid unnecessary encounters with neighbors.

◀ **Territorial boundaries** of a stable wolf pack may barely change over more than a decade. Where there is no natural barrier between territories, there may be a "no wolf's land" normally avoided by neighboring packs. During food shortages and as the pack eliminates more and more prey (eg deer) from its territory, the wolves are forced to seek prey in border areas, and more wolves are killed as pack encounters become more frequent. If deer numbers decline even further, the pack may risk trespass into or even through its neighbors' territory in search of prey. The departure of deer to winter yarding areas may also force a pack to risk trespassing into the territories that border the deer yards. When the wolf killed in a fatal encounter between packs is a dominant animal, the victim's companions may disperse, leaving a territory "up for grabs."

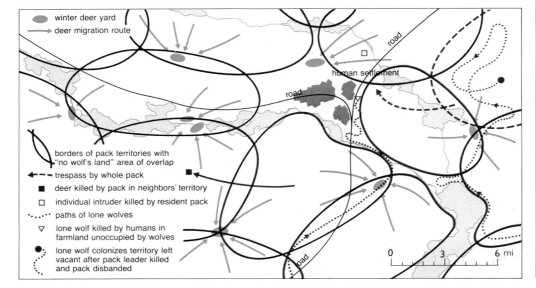

winter deer yard

deer migration route

borders of pack territories with "no wolf's land" area of overlap

trespass by whole pack

deer killed by pack in neighbors' territory

individual intruder killed by resident pack

paths of lone wolves

lone wolf killed by humans in farmland unoccupied by wolves

lone wolf colonizes territory left vacant after pack leader killed and pack disbanded

road

human settlement

road

road

0   3   6 mi

the need to defend a valuable resource appears to outweigh the relatively small risk of attack. Older pups, which can retreat with the pack, and already exploited kills are not sufficient cause to run this risk; reply rates of 30 percent have been observed in these conditions.

This resource-based decision is further modulated by pack size and season. A pack of 7–10 wolves replied on 67 percent of the nights when the observers howled to them, whereas a pack of 3–5 replied on only 40 percent; and during the breeding season, when interpack aggression reaches its zenith, reply rates increase for all packs.

There are two other reasons why failing to reply to strangers' howling may benefit a pack. If the pack desires to seek out its neighbors, it may be best to do so unannounced. And because a pack fails to reply about as frequently as it replies, neighbors are kept uncertain of the pack's whereabouts, and may refrain from entering an area despite their howls being unanswered.

A lone wolf keeps a lower profile than a pack. Loners—mostly younger animals that have left their natal pack—travel areas 10–20 times greater than does a pack. In this search for a place to settle, find a mate and start its own pack, the lone wolf rarely scent marks or howls. Many loners never reach their goal, but fall victim to hunters, trappers or hostile wolf packs. Once in possession of a vacant area, however, the lone wolf begins to scent mark and will howl readily in response to strangers, ready to defend its territory.　　　　FHH

less frequent than accidental ones, they confront the pack with the dilemma "to reply, or not to reply." If a pack seeks to avoid an encounter, it solves this dilemma by applying a simple rule. When the pack can do so with little loss, it usually slips silently away from the strangers. The silence offers no clues to strangers seeking a deliberate encounter, while any scent marks left by the pack during the retreat can help to prevent an encounter desired by neither pack. But, if to move off means that a pack risks losing an important resource, then it usually stays where it is and replies. The most important resources for the wolf pack are young pups and fresh prey kills: neither can be abandoned without a potentially great loss to the pack. Packs at fresh kills have replied to neighbors' howls in more than four out of every five cases observed. The pack's answering howls prevent any accidental meeting, and although they could assist strangers intent on an attack

COYOTE

Canis latrans
Coyote or Prairie wolf or Brush wolf.
One of 9 species of the genus *Canis*.
Family: Canidae.
Distribution: N Alaska to Costa Rica;
throughout Mexico, continental US and much
of W and C Canada.

Habitat: open country and grassland; may also
occupy deciduous, mixed coniferous and
mountain forests.

Size: head-body length 28–38in
(70–97cm); tail length 12–15in
(30–38cm); shoulder height
18–21in (45–53cm); weight
25–33lb (11.5–15kg).

Coat: grizzled buff-gray; muzzle, outerside of
ears, forelegs and feet dull brownish-yellow;
throat and belly white; prominent black stripe
down middle of back; black patches on front
forelegs and near base and tip of tail.

Gestation: 63 days.

Longevity: up to 14.5 years (18 in captivity).

▲ **A foraging coyote** follows intently the
progress of its prey, probably a small rodent or
insect.

► **The familiar howl** of the coyote, heard so
often in Western movies, is only one in an
elaborate repertoire of calls.

Tʜᴇ stark image of a solitary coyote is
familiar to many, thanks to its use in
countless Westerns. It has been a persistent
image, and for long the coyote was considered to be a solitary animal. But recent
studies have shown that in some situations
coyotes live cooperatively in a way similar to
wolves.

The coyote—whose name derives from
the original Aztec word for the species,
coyotl—is a medium-sized canid with a
rather narrow muzzle, large pointed ears
and long slender legs. Size varies between
populations and from one locale to another,
and adult males are usually heavier and
larger than adult females.

While the geographic ranges of most
predators are shrinking, that of the coyote is
increasing. A northerly and, particularly,
an easterly expansion from the central Great
Plains began in the late 19th century, as
local populations of the larger canids, the
Gray wolf (*Canis lupus*) and the Red wolf (*C.
rufus*), were decimated by man.

Coyotes can interbreed with the Domestic
dog, the Red wolf and, probably, the Gray
wolf (the so-called Eastern coyotes are now
thought to be fertile coyote–Gray wolf hybrids). The coyote–Domestic dog hybrid
("coydog") can reproduce at one year old
and has two litters a year. Coydogs are even
more liable to attack farm and domestic
animals than are coyotes.

Like jackals and wolves, the coyote is an
opportunistic predator. Mammals, including carrion, generally make up over 90
percent of its diet. Ground squirrels, rabbits
and mice predominate, but larger animals
such as the Pronghorn antelope, deer and
Rocky mountain sheep are included.
Coyotes also eat fruit and insects. Small prey
are hunted singly, but larger animals are
hunted cooperatively. Coyotes normally
stalk small prey from a few meters, but
occasionally from as far as 165ft (50m) and
for as long as 15 minutes. Two or more
coyotes may chase larger prey for up to
1,300ft (400m).

▲ **A solitary coyote** pauses while out hunting in Banff National Park, Alberta, Canada.

▲ **A coyote pack defends a carcass** on the edge of its territory. Three pack members (1) are feeding while the dominant male (2) aggressively threatens (tail bushy and almost horizontal, ears erect and slightly forward, fur erect on neck and shoulders, mouth open to expose canines) an intruder (3), who assumes a defensive threat posture (tail between legs, ears pressed back on head, mouth open to expose all teeth, back arched and hair erect along entire length of back and neck). Another male (4), backing up his dominant partner, shows less intense aggression. (5) Another trespasser watches for the outcome of the encounter. Other coyotes (6) wait in their own territory for the resident pack to leave the carcass.

The basic social unit in most coyote populations is the breeding pair; and the size of the home range varies from 5.5 to 25 sq mi (14 to 65 sq km) for males, with an average of 9.6 sq mi (25 sq km) for females. Coyotes are now known to form packs similar to wolf-packs, in certain situations. Such packs are formed by delayed dispersal of the young, who remain as "helpers" in a pack; a typical pack consists of about six closely related adults, yearlings and young. It is usually the dominant male and female that breed.

Pack members sleep, travel and hunt larger prey together and cooperate in territorial disputes and defense of carrion. In general, coyote packs are smaller than wolf packs and associations between individuals less stable. The reasons for this may be the early expression of aggression, which is found in coyotes but not in wolves, and the fact that coyotes often mature in their first year whereas wolves do so in their second.

Variation in social organization enables the coyote to thrive on diverse prey and this flexibility is probably the reason for the wide, and expanding, geographic range of the species. Coyotes living in packs are more effective predators of large animals, and where such prey (eg deer, elk) is available, packs of 3–8 coyotes are found. Where the principal prey is small mammals, the pups disperse early, packs are not formed, and most sightings are of solitary coyotes. Seasonal variation in social structure also occurs: when Ground squirrels and the young of large mammals are available as prey, coyotes spend less time together.

Coyotes use urine marking and calls to define their territory, to communicate with each other, and to strengthen social bonds. The coyote's howl is unique and consists of a series of high-pitched staccato yelps followed by a prolonged siren wail. Their vocalizations also include barks, bark-howls, group yip-howls and group howls.

During the last 150 years, coyotes have been responsible for large economic losses to US agriculture, especially sheep farming. Some ranchers have lost up to 67 percent of their lambs and 20 percent of their sheep to coyotes in a single year; others lose very few. In fact, there is evidence that attempts to control coyotes by poisoning may also deplete the numbers of their natural prey and lead to increasing attacks by coyotes on farm animals. Although the species is not endangered, it is now totally protected in 12 states and the coyote harvest is regulated by a hunting or trapping season in most of the remaining states and Canada.   WDB

Both sexes attain sexual maturity during the first breeding season (January to March) following birth. Females produce one litter a year, averaging six pups per litter. The young are born blind and helpless in a den and are nursed for a period of 5–7 weeks. At three weeks pups begin to eat semisolid food regurgitated by both parents and other pack members of both sexes. Most young disperse in their first year and may travel up to 100mi (160km) before settling down.

JACKALS

Four of 9 species of the genus *Canis*
Family: Canidae.
Distribution: Africa, SE Europe, S Asia to Burma.

Size: head-body length 26–42in (65–106cm); tail length 8–16in (20–41cm); shoulder height 15–20in (38–50cm); weight 15–33lb (7–15kg), averaging 24lb (11kg).

Gestation: 63 days.

Longevity: 8–9 years (to 16 years in captivity).

Golden jackal
Canis aureus
Golden or Common jackal.

N and E Africa, SE Europe, S Asia to Burma. Arid short grasslands. Coat: yellow to pale gold, brown-tipped.

Silverbacked jackal
Canis mesomelas
Silverbacked or Blackbacked jackal.

E and S Africa. Dry brush woodlands. Coat: russet, with brindled black-and-white saddle; fur finer than *C. aureus*.

Simien jackal [E]
Canis simensis
Simien or Ethiopian jackal, Simien fox.

Bale and Simien regions of Ethiopia. High mountains. Coat: bright reddish with white chest and belly.

Sidestriped jackal
Canis adustus

Tropical Africa. Moist woodlands. Coat: grayer than *C. aureus* with distinctive white stripe from elbow to hip and tail white-tipped.

[E] Endangered.

▶ **Mutual grooming** of these two adult Silverbacked jackals indicates their close family ties. Jackal family groups are tightly knit, and cubs are tended by parents and helpers of both sexes.

THE jackal has a bad name: the word can also mean "one who performs menial tasks for others, especially of a base nature." But the facts about jackals are rather more edifying than the popular image of a cowardly scavenger. Jackals are much less dependent on carrion than is commonly supposed, and their family life is noted for its stability: partnerships between male and female are unusually durable for a mammal.

Jackals are small slender dog-like omnivores with long sharp canines and well-developed carnassial teeth used for shearing tough skin. Like most other canids, jackals are lithe muscular runners with long legs and bushy tails. They have large erect ears and an elaborate repertoire of ear, muzzle and tail postures. The average weight of 24lb (11kg) applies to all species except the South African Silverbacked jackal, where in some populations the male is usually 2.2lb (1kg) heavier than the female.

Of the four species, the Golden jackal has the widest distribution (East Africa to Burma); the others are limited to Africa. In East Africa, distribution of Golden, Silverbacked and Sidestriped jackals overlaps; but each species occupies a different habitat. Skeletal remains of Silverbacked jackals in Bed I (1.7 million years old) of the Olduvai Gorge provide the earliest fossil record of a present-day *Canis*; the species still lives in the brush woodland nearby the gorge.

Coat color and markings also distinguish the species, the coat of the Simien jackal being the most colorful. In the Golden jackal, coat color varies with season and region: on the Serengeti Plain in north Tanzania it is brown-tipped yellow in the rainy season, changing to pale gold in the dry season.

Jackals are opportunistic foragers for their very varied diet. They eat fruits, invertebrates, reptiles, amphibia, birds, small mammals—from rodents to Thomson's gazelles—and carrion. In the Silverbacked jackal's diet, rodents and fruit are the most important items, and the fruit of the tree *Balanites aegyptiaca*, which is highly nutritious and often favored, may even function as a natural "wormer" to alleviate infestation by parasites. In the Serengeti, scavenging usually contributes less than 6 percent of a jackal's diet.

Cooperative hunting is important for Golden and Silverbacked jackals. In both species, pairs have been observed to be three times more successful than individuals in hunting Thomson's gazelle fawns. Members of the same family will also cooperate in sharing larger food items, ranging in size from hares to perhaps a wildebeest carcass, and will transport food in their stomachs for later regurgitation to pups or a lactating mother. Food is also cached.

Both sexes mature at 11 months, although they may not breed at once. Some yearlings stay on as "helpers" to assist their parents in raising the next litter. Serengeti Golden jackals court at the end of the dry

▲ **The four species of jackal** depicted in typical play postures. (**1**) Golden jackal cub attacking the twitching ear of an adult; (**2**) Silverbacked jackal juveniles in a tail-pulling game, the victim's arched back and laid-back ears revealing its unease; (**3**) two Simien jackal near-adults play-chasing; and (**4**) a Sidestriped jackal playing with a dead mouse.

▲ **A jackal pounces** onto a small animal. The steeply arched trajectory, stiff tail, forward ears and close-gathered front paws of this Silverbacked jackal characterize the hunting pounce of many canids.

▶ **The den of a jackal** may be a natural hole or crevice, a den built by another animal, or a new excavation – seldom more than 10ft (3m) long; sometimes jackals may simply hide in thickets. The present occupant of this rock-sheltered den is a Golden jackal.

Jackal Helpers

This typical Silverbacked jackal family group comprises the breeding female (**1**) and her pups (two shown suckling and one begging her to regurgitate food), and a "helper" (**2**) in submissive posture towards its father, the breeding male (**3**). Why some juveniles stay to live with their parents, apparently delaying their own reproduction by a year, and, instead, help raise their younger brothers and sisters is an intriguing question, and attempts to answer it have provided insight into how and why family bonds develop in a hunting and gathering mammal.

In many ways jackals have to contend with conditions not unlike those probably experienced by early man in Africa. In much the same environment, they, too, live in small close-knit families that share food and care for dependent young; and, when they mate, they probably mate for life. Mated pairs of Silverbacked jackals in the Serengeti have been observed to produce pups for at least six years; Golden jackals, for at least eight.

In a recent six-year study in the Serengeti, 12 of 19 litters of Silverbacked jackals, and 6 of 8 Golden jackal litters, had "helpers." Within the family, these helpers were subordinate to parents, and male and female alike assisted in guarding and feeding the pups.

When a jackal pair raises a litter unaided, pups will often be left on their own for up to 40 percent of the time—that is, after the first three weeks of virtually constant attention. But with a helper present, the pups are seldom left undefended. The presence of a single adult at the den provides considerable protection. Not only will an adult "rumble growl" or "predator bark," warning the pups to take refuge and threatening the predator, but a single adult can successfully drive off large predators. A jackal will not hesitate to chase and bite a hyena, even though the latter may be five times its own weight. Adult jackals are quick, and can close in and dart away before a hyena can turn and defend its rear.

Helpers also regularly bring food to the lactating mother, and there are more regurgitations of food, and more nursing sessions by the mother, in families with helpers. Helpers may also improve the provisioning of pups indirectly by allowing the parents to spend more time foraging alone or hunting as a pair. Also, larger families (families with helpers) may be able to defend and exploit a carcass more successfully than an individual. A single jackal, instead of feeding, may spend most of its time threatening vultures—and then often in vain.

Survival rates for Silverbacked jackals confirm the genetic advantage of helpers to members of this species. Silverbacked parents with no helpers raise on average only one pup; with just a single helper, three pups may survive, and a family with three helpers is known to have reared six pups. On average, the reproductive success of families observed increased by 1.7 pups with the addition of each helper. Moreover, since their parents are monogamous, jackal helpers are as closely related to their full siblings as they would be to their own offspring. In the short term, therefore, helpers can do more to ensure the survival of their own genes by caring for their brothers and sisters than they could by starting their own families.

There are of course other benefits to jackals from staying on the territory where they were born and extending their experience at home. They may live longer and may in time raise their own pups more successfully than they would otherwise. In a few cases helpers may actually inherit part of their parents' territory and benefit from long familiarity with it. The precise contribution of factors such as these to the evolution of helping behavior is, however, difficult to evaluate; and the fact that a one-year-old jackal can emigrate and yet still benefit genetically (more siblings survive), as long as one litter-mate does stay and help, is a typical complication.

Among Golden jackals, pup survival does seem to improve in the presence of helpers, but not as markedly as with Silverbacked jackals. Goldens whelp during the rainy season (December–January), when food on the shortgrass plains is most abundant. However, the rains that bring the wildebeest and zebra also flood dens, and pups die of exposure and illness. Thus, any long-term gains may, in the short term, be offset.

season (October) and produce pups in the rainy season (December–January). Thus whelping occurs during the period of greatest food abundance on the shortgrass plains. Silverbacked jackals whelp in July–October, which coincides with a peak in rodent numbers and with fruiting of the balanites tree. Most jackal species have a maximum litter size of nine, but a pregnant Sidestriped jackal has been found with 12 fetuses, so reabsorption of fetuses, or other early mortality, may occur. The number of pups which reach maturity varies; in Serengeti Silverbacked jackal litters without a helper, only one pup survives on average; with one helper, three or more may live.

Evidence of helping behavior has emerged in all members of the dog family that have been studied in detail. The jackal helper (see LEFT) is fully mature (jackals reach sexual maturity at 11 months), and while subordinate to its parents, is an important member of the family.

Young jackal pups remain in the den—a natural shelter or simple excavation—for some three weeks, during which time the mother spends about 90 percent of her time with them, perhaps to keep them warm. She nurses the pups until they are about eight weeks old. The female initiates all den changes, which may be as frequent as every two weeks. When Silverbacked jackal pups are three months old they stop using a den. Pups are fed regularly by regurgitation until about five months old, and may occasionally be fed until they leave their parents, at one or two years of age.

Jackal pairs hold territories of 0.2–1 sq mi (0.5–2.5 sq km) throughout the year. They forage and rest together and all their behavior is highly synchronized. They tend to scent mark their territory in tandem, with either male or female making the first mark, probably advertising to intruders that both members of the pair are in residence. The males are strictly monogamous, perhaps because divided paternal care might reduce the number of cubs that survive. Females reserve their aggression for female intruders, thus preventing the sharing of the male and his paternal investment. Since females may be able to produce litters with several sires, the male needs to ensure that he alone copulates with his mate, if a wasted investment in pups that he has not sired is to be avoided. It is not surprising therefore that the male of a pair fiercely threatens and attacks any male intruder on the territory. Both members of a pair have important roles in maintaining their territory and in raising the young. When one parent dies, the rest of the family is unlikely to survive.

In addition to posture and scent, each jackal species communicates through its own repertoire of calls. These may include howls, yelps, barks and other vocalizations. Silverbacked and Golden are more vocal than Sidestriped jackals. In the Serengeti Goldens locate each other by howling, while the contact call of the Silverbacked jackal is a series of yelps.

In agricultural areas jackals tend to be persecuted by man; and jackals are also killed for their fur. In Ethiopia the Simien jackal's numbers have been so reduced that this is now an endangered species, with only an estimated 500 left in the wild.   PDM

▼ **Carrion,** though important, forms a very much smaller part of a jackal's diet than has been traditionally supposed. This group of Silverbacked jackals deters vultures which would almost certainly succeed in driving away a solitary jackal.

FOXES

Twenty-one species in 4 genera
Family: Canidae.
Distribution: Americas, Europe, Asia, Africa.

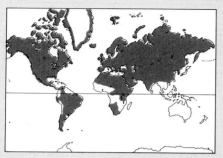

Habitat: very wide-ranging, from Arctic tundra to city center.

Size: ranges from head-body length 9.5–16in (24–41cm), tail length 7–12in (18–31cm) and weight 3.3lb (1.5kg) in the Fennec fox to head-body length 28–39in (72–100cm), tail length 10–14in (25–35cm) and weight 20lb (9kg) in the Small-eared dog.

Gestation: between 50 days (Fennec fox and others), and 60–63 days (Red fox).

Longevity: up to 6 years (13 in captivity).

Vulpine foxes

Twelve species of the genus *Vulpes*.
N and S America, Europe, Asia, Africa.
Coat: mostly grays to red-browns, but white, silver, cream, buff-yellow and black also.
Species include: the **Red fox** (*V. vulpes*) or Silver or Cross fox; the **Gray fox** (*V. cinereoargenteus*) or Tree fox; the **Swift fox** (*V. velox*) or Kit fox; the **Fennec fox** (*V. zerda*); and the **Indian fox** (*V. bengalensis*).

South American foxes

Seven species of the genus *Dusicyon*.
Coat: mostly grays with tawny grizzling, often with black markings.
Species include: the **Argentine gray fox** [*] (*D. griseus*); the **Colpeo fox** [*] (*D. culpaeus*); and the so-called **Small-eared dog** [?] (*D. microtis*).

Arctic fox

Alopex lagopus
Arctic or Polar or White or Blue fox.
N polar regions.
Coat: very thick, either white (gray-brown in summer) or blue (chocolate-brown in summer)

Bat-eared fox

Otocyon megalotis
Bat-eared fox or Delandi's fox.
E and S Africa.
Coat: gray to buff; face markings, tips of ears, feet and dorsal stripe all black.

[*] CITES listed. [?] Threat suspected.

IN Aesop's fable the cunning fox outwits the stork; and in the Uncle Remus stories Br'er Fox is the chosen adversary of Br'er Rabbit. That foxes figure as the wily character in the popular tales of many lands reflects both their distribution and their resourceful behavior. The largest genus of foxes, *Vulpes*, is also the most widespread of canid genera, and one of its members, the Red fox, is the most widely dispersed and arguably the most adaptable of all carnivores.

Foxes are small canids with pointed muzzles, somewhat flattened slender skulls, large ears and long bushy tails. All species are opportunistic foragers, using hunting techniques which vary from stealth to dash-and-grab. Apart from the Bat-eared fox (see p65) which eats mainly termites, there are no proven differences in species' diets other than those imposed by the limits of available prey. Arctic foxes will take sea birds, ptarmigan, shore invertebrates, fruits and berries, together with carrion found while methodically beachcombing. They time their shore visits to coincide with the receding tide when fresh debris is stranded. Red foxes have similarly diverse diets, ranging from small hoofed mammals, rabbits, hares, rodents and birds to invertebrates such as beetles, grasshoppers and earthworms. Red foxes have been observed to fish, wading stealthily through shallow marshes. In season, fruit such as blackberries, apples and the hips of the Dog rose can form as much as 90 percent of the diet.

All vulpine foxes catch rodents with a characteristic "mouse leap," springing a meter off the ground and diving, front paws first, onto the prey. This aerial descent may be a device literally to squash the vertical jump used by some mice to escape predators. Red foxes catch earthworms that leave their burrows on warm moist nights by crisscrossing pastures at a slow walk and listening for the rasping of the worms' bristles on the grass. Once a worm is detected, the fox poises over it before plunging its snout into the grass. Worms whose tails retain a grip in their burrows are not broken but gently pulled taut after a momentary pause—a technique foxes share with bait-collecting fishermen.

The few fox species that have been studied in several different habitats have been found to eat whatever food was locally available. Nevertheless, foxes may have some preferences. For example Red foxes, if given the choice, will prefer rodents of the family Microtinae, such as field voles, to members of the family Muridae, such as field mice. However, being true opportunists, they will cache even unfavored prey for future use, and have a good memory for the location of these larders.

The similar foraging behavior of different fox species may affect their geographical distribution, as it leads to severe competition for food. Arctic and Red foxes were once thought to be separated by the Arctic species' remarkable tolerance to cold temperatures—its metabolic rate does not even start to increase until −58°F, in contrast to the Red fox's, which increases at 8.6°F. However, Red foxes are sometimes found in even colder places than Arctic foxes, so they are probably separated by food

▲ **Unmistakably vulpine.** the mask of the Red fox, with its prominent ears, eyes and whiskers, fine muzzle and sensitive nose, epitomizes the species' alertness – often unfairly portrayed in fables as sly cunning.

▶ **The coat of the northern Red fox** occurs in three color forms, here represented in a group of foxes scavenging from a carcass in a northern forest. (**1**) Two individuals with the vivid flame-red coloring typical of most high latitude Red foxes; (**2**) the melanistic ("Silver") form; and (**3**) the intermediate so-called Cross fox. The different forms are probably under complex control of two different genes.

◀ **South American foxes** of the genus *Dusicyon.* (**1**) the Small-eared dog (*D. microtis*): (**2**) the Colpeo fox (*D. culpaeus*); (**3**) the Argentine gray fox (*D. griseus*); (**4**) Azara's fox (*D. gymnocercus*); and (**5**) the Crab-eating fox (*D. thous*).

▲ **Smallest of all canids,** the Fennec fox, fully grown, weighs scarcely 3.3lb (1.5kg). With its large ears and pale fur, it is well adapted to the desert conditions of northern Africa, Sinai and Arabia where it lives.

▼ **Eight vulpine species,** depicted in a dash and swipe after a bird, shown in west-to-east order of distribution (only Gray and Corsac, and occasionally Red foxes climb trees). (1) Gray fox (*Vulpes cinereoargenteus*); (2) Swift fox (*V. velox*); (3) Cape fox (*V. chama*); (4) Fennec fox (*V. zerda*); (5) Rüppell's fox (*V. rüppelli*); (6) Blanford's fox (*V. cana*); (7) Indian fox (*V. bengalensis*); (8) Corsac fox (*V. corsac*).

competition: the Red is up to twice as heavy, needs correspondingly more food, and thus in the far north, where prey is sparse, cannot match energy gains with expenditure in the way that the Arctic fox can. However, in areas where both species can subsist, the Red fox's greater size enables it to intimidate the Arctic fox and in effect determine the southern limit of the latter's range. The ranges of these species doubtless fluctuated as Ice Ages came and went.

Direct competition may also have affected the distribution and sizes of *Dusicyon* species. In central and southern Chile both the Colpeo fox and the Argentine gray fox eat rodents, birds, birds' eggs and snakes in comparable quantities. However, these two species vary in size with latitude throughout their range. Average body length of the Colpeo fox increases from 28–35in (70–90cm) and that of the Argentine gray decreases from 27–17in (68–42cm) with change in latitude from 34°S to 54°S. Where the two species are of similar size (34°S), the Colpeo inhabits higher altitudes of the Andes, so reducing competition. Further south, where the altitude of the Andes decreases, bringing the species into apparently direct competition, the much smaller size of the Argentine gray fox predisposes it to hunting smaller prey than the Colpeo fox, and so, again, competition is reduced.

Foxes breed once a year. Litter sizes are normally from one to six, the average for the Red fox varying with habitat between four and eight. The maximum number of fetuses found in a Red fox vixen is 12. Vixens have six mammae, or teats. Known gestation periods are 60–63 days for the Red fox and 51 for the Fennec fox. Cubs are generally born in burrows (either dug by the vixen or appropriated from other species) or rock crevices. Litters of Red foxes have been found in hollow trees, under houses or simply in long grass. Foxes have generally been considered as monogamous, but communal denning has now been recorded for the Indian fox and the Red fox, and "helpers" at the den occur in both Arctic

and Red foxes (see below). Among Red foxes the proportion of vixens which breed varies greatly between areas, from 30 percent to almost all.

Foxes have been characterized as solitary carnivores, foraging alone for small prey for which cooperative hunting would be a hindrance rather than an advantage. In this respect their social behavior has been contrasted with that of pack-hunting canids such as wolves. However, with modern radiotracking studies and with night-vision equipment, it has become clear that fox society is complex. In some areas foxes are monogamous; in others, as with Red and Arctic foxes, they may live in groups, generally composed of one adult male and several vixens. So far, the maximum proven adult group size for Arctic foxes is three and for Red foxes six. There is no evidence of successful immigration of vixens into such groups, so the female members are probably relatives, whereas all male offspring emigrate. Dispersal distance varies with habitat and records of over 125mi (200km) exist. Males invariably disperse farther than females.

Although their paths may cross many times each night, foxes within a group may

▲ **Gray fox in a tree.** Not surprisingly this tree-climbing species of the Americas is also known as the Tree fox. Although tree climbing is most common in Gray foxes, Red foxes also occasionally climb – in some places even sleeping in bushy trees.

◄ **Recently reclassified** as a subspecies of the Swift fox, the prairie-dwelling Kit fox of the southwestern United States and northwestern Mexico is smaller yet has larger ears than other North American foxes. Eight of the 10 subspecies of Swift fox share the common name Kit fox.

that of strangers. Foxes have paired anal sacs on either side of the anus. These can be evacuated voluntarily, or the secretion may be coated onto feces. Foxes also have a skin gland, 0.8in (2cm) long, on the dorsal surface of the tail, near the base. This "supracaudal gland," or "violet gland," is covered in bristles and appears as a black spot on the tails of all vulpine foxes. Its function is unknown. There are yet more glands between their toes. Both males and females may "cock their legs" when urine marking.

Territory sizes are probably determined by the availability of food and by the mortality rate, which is mainly dependent on man and rabies. Where mortality through hunting is high few Red foxes survive three years. The oldest vixen known from the wild was the nine-year-old matriarch of a group of four occupying a 100-acre (40-hectare) territory in Oxfordshire, England.

Like other canids, foxes communicate by means of sound as well as by means of scent marking and postural signals. The Arctic fox calls quite often—for example when an enemy approaches, or during the breeding season. The vocalizations of the Red fox include aggressive yapping and a resonant howl used by young foxes in winter and, more often, in the mating season. Barks, soft whimpers (between vixen and pups) and screams are also part of the repertoire of the Red fox.

Despite their fabled cunning, foxes, too, have their endangered species. Both the little-known Small-eared dog of tropical South America and a subspecies of the Swift fox, the Northern Swift fox (*V. velox hebes*), are listed as threatened by the IUCN. The latter, a small 4.4lb (2kg) North American prairie-dwelling fox, was rarely seen in the northern Great Plains between 1900 and 1970, and seems to have been completely

forage mainly in different parts of the territory, with dominant animals monopolizing the best habitat.

Range sizes for Red foxes have been found to vary between 25 and 5,000 or more acres (10 and 2,000 hectares), those of Arctic foxes between 2,100 and 15,000 acres (860 to 6,000 hectares). Territory area and group size are unrelated. Feces and urine are left on conspicuous landmarks, like tussocks of grass. These scent marks are distributed throughout the foxes' range, but especially in places visited often. Dominant animals scent mark with urine more than subordinates do, and individuals can distinguish the scent of their own urine from

5   6   7   8

◀ **White form of the Arctic fox** in its smoky gray summer coat. The short, rounded ears and short muzzle are heat-saving adaptations to the polar climate, as is the fur that lines the soles of its feet. Arctic foxes are often observed following Polar bears in order to scavenge from their kills.

▶ **The coat of Arctic foxes** occurs in two color forms, each with a different summer and winter coloration. (**1**) A white, or Polar, form in winter coat attacks a ptarmigan; (**2**) the same form in summer coat; (**3**) a steel-gray to brown, or Blue, form beachcombs for carrion, again in winter; and (**4**) a Blue fox vixen "helper" in brown summer coat with her year-younger siblings. This litter contains both color forms – the light gray cubs have the summer coat of the white form, while the dark gray cub is a further variant of the summer coat of the Blue form. The two color forms are genetically determined by a single gene, the white form being recessive. The ratio of white to blue varies: in mainland Alaska, Canada and Eurasia less than 1 percent are blue, but on small islands like the Pribilofs more than 90 percent are blue. The extent of snow cover in the habitat seems to influence which form predominates.

▼ **Very large ears** of these Bat-eared foxes leave little doubt as to why the species is so named. The ears are an adaptation to locating insect prey by sound and also assist heat loss in their hot environment.

exterminated by hunting in Canada. In 1928 it disappeared from Saskatchewan and in 1938 from Alberta. However, Swift foxes seem to have reappeared recently within the range previously occupied by the Northern Swift fox, although they may be the Southern form, *V. v. velox*.

The management and conservation of foxes turn on three main issues: killing for sport or predator control; hunting for pelts; and rabies control. The past decade has witnessed a revival of the prewar vogue for fox skins and has resulted in huge harvests. In the 1977–78 season, 388,643 Red foxes, 264,957 Gray foxes and 37,494 Arctic foxes were taken in North America. In 1978, nearly one million skins of the Argentine gray fox were reputedly exported from Argentina. The risk of over-harvesting has prompted several of the midwestern states of the United States to limit fur licenses severely and to extend closed seasons. Outside North America there is little control of the trade in canid skins. The new market for pelts has transformed Britain and Ireland, where there are no monitoring systems, into fur-trapping nations for the first time in recent history, and it is estimated that 50,000 to 100,000 Red fox skins were exported in 1980, with prices in 1979 averaging $60.

All genera of foxes are liable to spread,

and to suffer from, rabies, the virus disease that causes hydrophobia in humans. Millions of foxes have been slaughtered in unsuccessful attempts to control the disease, but foxes have such resilience that populations can withstand about 75 percent mortality without further declining. The best hope for eliminating rabies may lie in oral vaccination—accomplished by air-dropping fox baits containing anti-rabies vaccine. Preliminary trials in Switzerland and Canada have shown that up to 74 percent of foxes will eat the bait. However, an effective killed vaccine suitable for oral administration has yet to be perfected.

DWM

The Bat-eared Fox—An Insect Eater

Besides the ears (among foxes only the Fennec fox's are larger in relation to its body), it is teeth and diet that set this African species apart from all other foxes. Although usually found near large herds of hoofed mammals, such as zebras, wildebeest and buffalo, the Bat-eared fox is the only canid to have largely abandoned mammalian prey.

The teeth of the Bat-eared fox are relatively small, but it has between four and eight extra molars (see p49). On the lower jaw a step-like protrusion anchors the large digastric muscle that is used for rapid chewing. The diet includes fruits, scorpions and the occasional mammal or bird, but 80 percent is insects, and of these most are termites, particularly the Harvester termite (*Hodotermes*) and dung beetles (Scarabidae).

These insects abound only where large ungulates are numerous—the Bat-eared fox remains dependent on mammals for provision of its food, despite the insectivorous nature of its diet. Colonies of Harvester termites (which can make up 70 percent of the fox's diet) live underground, but large parties surface to forage for small blades of grass, from which they are taken by the foxes, and such grass occurs mainly where large ungulates feed. Dung beetles eat the dung of ungulates and lay their eggs in dung balls, which the female beetles bury. Bat-eared foxes eat both adults and larvae, which they locate by listening for the sound of the grub as it gnaws its way out of the dung ball. In this, the large ears, up to 5in (12cm) long, serve the fox well.

Bat-eared foxes usually breed in self-dug dens in pairs, but there are records of a male with two breeding females (and one record of communal nursing). The pups (usually 2–5) are born after a 60-day gestation. Juveniles achieve full adult size at four months and, when accompanying adults, may account for reported group sizes of up to 12. Foraging is usually done on an individual basis.

Extensive overlap of foraging areas has been reported. Interaction between groups varies from peaceful mingling to overt aggression. Two or three breeding dens are sometimes clustered within a few hundred meters—probably a response to locally suitable soil or vegetation. Reported home range sizes vary from 0.2–1.1sq mi (0.5–3sq km). Population density may reach 26 per square mile (10/sq km), but 1.3–7.8 per square mile (0.5–3/sq km) is more usual.   JM

THE 21 SPECIES OF FOXES

Abbreviations: HBL = head-body length; TL = tail length; wt = weight.
[*] CITES listed. [?] Threat to species suspected.

Genus *Vulpes*

N and far north of S America, Europe, Africa and Asia. Muzzle pointed; ears triangular and erect; tail long and bushy; skull flattened, by comparison with *Canis*. Tail tip often different color from rest of coat (eg black or white); black triangular face marks between eyes and nose.

Indian fox
Vulpes bengalensis
Indian or Bengal fox.

India, Pakistan and Nepal. Steppe, open forest, thorny scrub and semidesert up to altitudes of 4,400ft. Regarded anatomically as the "typical" vulpine fox. HBL 18–24in; TL 10–14in; wt 4–7lb. Coat: sandy-orange with legs tawny-brown; tail black-tipped.

Blanford's fox [*]
Vulpes cana
Blanford's or Hoary or Baluchistan fox.

Afghanistan, SW USSR, Turkestan, NE Iran, Baluchistan. Mountainous regions. HBL 16in; TL 12in; wt 7lb. Coat: as Indian fox, but blotchy with dark mid-dorsal line and brown chin.

Cape fox
Vulpes chama
Cape or Silverbacked fox.

Africa S of Zimbabwe and Angola. Steppe, rocky desert. Anatomy, especially skull, similar to Indian fox and Pale fox. HBL 18–24in; TL 12–16in; wt 8–10lb. Coat: rufous-agouti with silvery-gray back; tail tip black; dark facial mask lacking; ears elongate.

Gray fox
Vulpes cinereoargenteus
Gray or Tree fox.

Central USA to the prairies, S to Venezuela, N to Ontario. Able to climb trees. Formerly separated as *Urocyon*, partly because of longer tail gland. HBL 20–27in; TL 11–18in; wt 5–15lb. Coat: gray-agouti; throat white; legs and feet tawny; mane of black-tipped bristles along dorsal surface of tail. Canine teeth shorter than average for *Vulpes*.

Corsac fox
Vulpes corsac

SE USSR, Soviet and Chinese Turkestan, Mongolia, Transbaikalia to N Manchuria and N Afghanistan. Steppe, HBL 20–24in; TL 9–14in; wt unknown. Coat: russet-gray with white chin. Subspecies: 3.

Tibetan fox
Vulpes ferrilata
Tibetan or Tibetan sand fox.

Tibet and Nepal. High steppe (14,700–15,700ft). Although the most *Dusicyon*-like of *Vulpes* species, probably descended from Corsac fox. HBL 26in; TL 11in; wt not known. Coat: pale gray-agouti on body and ears; tip of tail white. Head long and narrow; canines elongate.

Island gray fox
Vulpes littoralis

Islands of W USA. Smaller than Gray fox but otherwise identical. HBL 23–31in; TL 4–11in; wt not known.

Pale fox
Vulpes pallida

N Africa from Red Sea to Atlantic, Senegal to Sudan and Somalia. Desert. HBL 16–18in; TL 11–12in; wt 6lb. Coat: pale fawn on body and ears; legs rufous; tail tip black; no facial marks; fur short and thin; whiskers relatively long and black.

Rüppell's fox
Vulpes ruppelli
Rüppell's or Sand fox.

Scattered populations between Morocco and Afghanistan, NE Nigeria, N Cameroun, Chad, Central African Republic, Gabon, Congo, Somalia, Sudan, Egypt, Sinai, Arabia. Desert. HBL 16–20in; TL 10–14in; wt 4lb. Coat: pale, sandy color; conspicuous white tail tip and black muzzle patches; whiskers relatively long and black. Subspecies: 6.

Swift fox
Vulpes velox
Swift or Kit fox.

NW Mexico and SW USA, N through prairie states to Alberta, Canada. Arid steppe and prairies. Kit foxes (8 of subspecies) formerly classified as *V. macrotis*. HBL 15–20in; TL 9–12in; wt 4–7lb. Coat: buff-yellow; limbs and feet tawny; black tip to very bushy tail. Subspecies: 10.

Red fox
Vulpes vulpes
Red or Silver or Cross fox.

N Hemisphere from Arctic Circle to N African and C American deserts and Asiatic steppes. Wide-ranging: Arctic tundra to European city centers. Natural S limit in Sudan. Introduced into Australia. (The N American subspecies *V. v. fulva* was formerly considered a separate species *V. fulva*.) Male HBL 27in; TL 17in; wt 13lb. Female HBL 26in; TL 16in; wt 11lb. (*V. v. fulva*: HBL 22–24in; TL 14–16in; wt 9–12lb.) Coat: rust to flame-red above (with silver, cross and color phases); white to black below; tip of tail often white. Subspecies: 48.

Fennec fox
Vulpes zerda

N Africa, throughout Sahara, E to Sinai and Arabia. Sandy desert. Formerly separated as *Fennecus* because of large ears, rounded skull and weak dentition. HBL 9–16in; TL 7–12in; wt 2–3lb. Coat: cream with black-tipped tail; soles of feet furred; ears very large, up to 6in long; dark bristles over tail gland; whiskers relatively long and black. The smallest fox.

Genus *Dusicyon*

Restricted to S America. Anatomy intermediate between *Vulpes* and *Canis*, with the extinct *D. australis* (Falkland Island wolf) most dog-like and *D. vetulus* most vulpine. Coat: usually gray with tawny grizzling. Skull long and thin; ears large and erect; tail bushy. Biology poorly known.

Colpeo fox [*]
Dusicyon culpaeus.

Andes from Ecuador and Peru to Tierra del Fuego. Mountains and pampas. (*D. culpaeolus* found in Uruguay is similar, but smaller, whereas *D. inca* from Peru is larger; both may be better considered subspecies of *D. culpaeus*, as indeed may be Azara's fox.) HBL 24–45in; TL 12–18in (see p62); wt 3lb. Coat: back, shoulders grizzled gray; head, neck, ears, legs tawny; tail black-tipped. Subspecies: 6.

Argentine gray fox [*]
Dusicyon griseus
Argentine gray or Gray or Pampas fox.

Distribution as Colpeo fox but lower altitudes in Ecuador and N Chile. Plains, pampas and low mountains. HBL average 16–27in (see p70); TL 12–14in; wt 10lb. Coat: brindled pale gray; underparts pale. Subspecies: 7.

Azara's fox
Dusicyon gymnocercus
Azara's or Pampas fox.

Paraguay, SE Brazil, S through E Argentina to Rio Negro. Pampas. Perhaps same species as Colpeo fox. HBL 24in; TL 13in; wt 10–14lb. Coat: uniform grizzled gray. Subspecies: 2.

Small-eared dog [?]
Dusicyon microtis
Small-eared dog, Zorro negro, Small-eared zorro.

Amazon and Orinoco basins, parts of Peru, Colombia, Ecuador, Venezuela, Brazil. Tropical forests. Formerly considered sufficiently distinct to be placed in separate genus *Atelocynus*. Ears short (2in) and rounded; has long, heavy teeth and enlarged second lower molar; gait reputedly cat-like. HBL 28–39in; TL 10–14in; wt 20lb. Coat: dark.

Sechuran fox
Dusicyon sechurae

N Peru and S Ecuador. Coastal desert. HBL 21–23in; TL about 10in; wt about 10lb. Coat: pale agouti without any russet tinges; tail black-tipped.

Crab-eating fox
Dusicyon thous
Crab-eating fox or Common zorro.

Colombia and Venezuela to N Argentina and Paraguay. Savanna, llanos and woodland. Formerly considered sufficiently distinct to be placed in separate genus *Cerdocyon*. HBL 24–27in; TL 11–12in; wt 13lb. Coat: gray-brown; ears dark; tail with dark dorsal stripe and black tip; foot pads large; muzzle short. Subspecies: 7.

Hoary fox
Dusicyon vetulus
Hoary fox or Small-toothed dog.

S central Brazil: Minas Gerais, Matto Grosso. Previously allocated to *Lycalopex* on basis of small teeth. HBL 24in; TL 12in; wt 6–9lb. Dark line present on dorsal surface of tail. A small *Dusicyon*, with short muzzle, small teeth and reduced upper carnassials.

► **Crab-eating fox,** with the grizzled gray coat, long pointed skull and large ears typical of the South American genus *Dusicyon*.

Genus *Alopex*

Similar to *Vulpes* but distinguished on basis of skull characters and adaptations to polar climate.

Arctic fox
Alopex lagopus
Arctic or Polar or Blue or White fox.

Circumpolar in tundra latitudes. Tundra and intertidal zone of seashore. Male: HBL 22in; TL 12in; wt 8lb. Female: HBL 21in; TL 12in; wt 7lb. Coat: very thick with two dichromatic color forms: "white" which is gray-brown in summer; and "blue" which is chocolate-brown in summer; 70% of fur is fine, warm underfur. Has remarkable tolerance to cold and only starts shivering at −94°F (−70°C). Subspecies: 9.

Genus *Otocyon*

Distinguished from other genera on the basis of unusual dentition.

Bat-eared fox
Otocyon megalotis
Bat-eared fox or Delandi's fox.

Two populations, one from S Zambia to S Africa, the other from Ethiopia to Tanzania. Open grasslands. HBL 18–23in; TL 9–13in; wt 7–10lb. Coat: gray to buff, with face markings, tips of ears, feet and dorsal stripe all black. Ears large (to 12cm); teeth weak but with extra molars to give 46–50 teeth in all; lower jaw with step-like angular process to allow attachment of large digastric muscle used for rapid chewing of termites. Subspecies: 2.

Fox Classification

The classification of foxes is complicated by similarities between imperfectly distinguished species and by fragmentary knowledge of their behavior. The arrangement adopted here is already contracted, but even so it could probably be further simplified. The genera are mainly distinguished by the fact that the "brows" formed by the frontal bones above and between the eyes are slightly indented or dished in the genus *Vulpes* and flat in *Dusicyon* (as opposed to convex in *Canis*). Familiarity with the widespread Red fox has lured some people into considering it as the "typical" vulpine fox, whereas it is perhaps the least typical—for example in being the largest. On the basis of morphology, the Indian fox is the "average" *Vulpes*, and even the Gray fox and the Fennec fox—species once placed in the separate genera *Urocyon* and *Fennecus*—conform to that average more closely than does the Red fox. Similarly, the South American Azara's fox and the Argentine gray fox are typical of *Dusicyon*; and the Small-eared dog (once classified as *Atelocynus microtus*) and the Crab-eating fox (once *Cerdocyon thous*) are so similar to them that to split them off from *Dusicyon* seems an unnecessary complication.

In both *Vulpes* and *Dusicyon* some geographically close species may be related, as in the case of the mountain-dwelling Tibetan fox and the similar but larger Corsac fox of lower altitudes in central Asia, from which the former is probably descended. In contrast, the Cape fox of southern Africa is morphologically most similar to the geographically distant Indian fox, which inhabits similar steppe country. Those species, which in a north–south arc separate the Indian fox and the Cape fox, namely Rüppel's fox, the Fennec fox and the Pale fox, are close relatives whose differences are adaptations to various arid landscapes. The distribution of Blanford's fox falls within this arc, but its small size—head-body length only 16.5in (42cm)—is probably an adaptation to a mountainous rather than a desert existence.

The desert foxes are the smallest, the lightest-colored and have the largest whiskers and external ears. This trend culminates in the very small 3.3lb (1.5kg) Fennec fox, whose heavily furred paws and large ears are adapted to life on hot sand dunes. Another fox specifically adapted to an extreme environment is the Arctic fox, whose species name (*Alopex lagopus*), indicates the resemblance of the long fur of the paws to that of a hare. Although very similar to, and sometimes classified within, *Vulpes*, the Arctic fox is sufficiently distinct (for example, in coloration) to merit generic status. Recent field studies have highlighted considerable similarities in the behavior of Red and Arctic foxes, and the two probably share a common ancestor, *Vulpes alopecoides*, from which they diverged in the Pleistocene Ice Ages, 1–2 million years ago.

Like the Arctic fox, the Bat-eared fox of Africa is placed in its own genus. *Otocyon* is separated on the basis of its dentition, unique among canids, which has evolved as an adaptation to an insectivorous diet.

The South American *Dusicyon* foxes are intermediate in appearance between the popular images of foxes and that of canids such as wolves and coyotes; indeed, in 1868 the taxonomist J. E. Gray called them Fox-tailed wolves. The Falkland Island wolf (*D. australis*) became extinct in 1880. It was first described by Darwin, who noted that it fed on the goose *Chloephaga picta*. In their skull and teeth these wolf-like foxes were more similar to *Canis* than *Dusicyon*, but their extermination by pelt-hunters has sadly relegated them to the status of a biological mystery. Some maintain that *D. australis* was descended from a domestic canid and not from a *Dusicyon* ancestor, pointing out that it has a white-tipped tail in contrast to the black one of all surviving *Dusicyon* species. Another South American mystery is *D. hagenbecki*, a fox-like canid known from a single skin collected in the Andes.

AFRICAN WILD DOG

Lycaon pictus [E]
African wild dog or Cape hunting dog or
Tri-colored dog or Wild dog.
Sole member of genus.
Family: Canidae.
Distribution: from the Sahara to S Africa.

Habitat: from semidesert to alpine; savanna
woodland probably preferred.

Size: head-body length 30–40in
(75–100cm); tail length 12–16in
(30–40cm); shoulder height 30in
(75cm); weight 44–60lb
(20–27kg). No variation between
sexes and little variation between
populations.

Coat: short, dark with a unique pattern of
irregular white and yellow blotches on each
individual; muzzle dark; tail white-tipped.

Gestation: 70–73 days.

Longevity: about 10 years.

[E] Endangered.

▼ **Impressive shearing teeth** displayed, as a
male African wild dog interrupts his midday
rest to yawn and to urinate: the posture
indicates a subordinate status in the pack of
which he is a member – only the dominant
male and female cock a hind leg while
urinating.

Aftera chase of some two miles, a
gazelle suddenly swerves. The African
wild dog at its heels fails to follow the jink,
but the second dog a few meters behind has
to run only one side of a triangle to the
gazelle's two, and is able to intercept and
seize the tiring prey. In the face of a cooperative hunter like the African wild dog,
the evasive action that would have foiled a
solitary cheetah seals the gazelle's fate.

The African wild dog is the least typical
canid. It is exclusively carnivorous and has
a short powerful muzzle that houses an
impressive array of shearing teeth, but with
the last molar poorly developed; unlike
other canids it has lost the fifth digit on the
front feet. The large rounded ears are used in
signals to other wild dogs and in controlling
body temperature.

Although savanna woodland is probably
the preferred habitat, wild dogs have been
recorded in the Sahara and in the snows of
Mount Kilimanjaro at 18,500ft (5,600m).
They are seldom numerous and densities as
low as one pack per 770sq mi (2,000sq km)
are not uncommon.

African wild dogs hunt, rest, travel and
reproduce in packs which average 7 or 8
adults but vary from 2 to 20 in number.
With pups a pack may contain over 30
individuals. Packs have a home base only
when young pups are present. At other
times the pack crisscrosses its range in the
course of seeking and catching prey, covering from 1–30mi (2–50km) each day. The
home ranges of packs studied on the
Serengeti Plain (north Tanzania) average
580sq mi (1,500sq km). The ranges of
different packs overlap extensively. If packs

meet, the larger group usually chases away
the smaller.

Wild dogs rely mainly on sight when
hunting, which usually takes place in the
cool hours around dawn and dusk, only
occasionally at night. Packs hunt at least
once a day, and large packs, particularly
when taking smaller prey, at least twice.

The chief prey species varies, with impala
in southern Africa, puku in parts of Zambia,
and Thomson's gazelle in most of East
Africa. All of these weigh between 44 and
200lb (20–90kg), but herbivores as small as
Cane rats (11lb/5kg) and as large as Greater
kudu (about 680lb/310kg) have been reported in the diet.

Hunting success depends on the ability to
choose young or weak animals, often selected from among a herd, that can be
overhauled in a straight chase. African wild
dogs can run at about 37mph (60km/h) for
3mi (5km) or more. Normally all adults
present share the kill, but adults will stand
aside to let any pups present eat first.

The adult males and adult females in a
pack have separate dominance hierarchies.
The dominant male and female usually stay
close to each other. About once a year—
often at a time of relative food abundance,
the dominant female is mated by the dominant male. The female selects a den, usually
a preexisting hole, and pups are born blind
and weighing about 14oz (400g). At about
three weeks of age their eyes open, and the
pups emerge above ground when they start
to eat solid food regurgitated by all the
adults. Once they are weaned, at about 10
weeks, the pups rely on food (and protection) provided jointly by the pack until they
can fend for themselves at about 14 months.
When the pups are about three months old,
they start to roam with the pack. Sexual
maturity is attained between 12 and 18
months.

Most males born in a pack will remain
there through their 10 years or so of life.
Females between 14 and 30 months of age
leave their natal pack in groups of littermate
sisters, and emigrate to a different male kin
line (see pp70–71).

The African wild dog is listed by the IUCN
as threatened by extinction and probably
the world population does not exceed
10,000. They have been persecuted by man,
are susceptible to epidemic diseases, such as
distemper, and their habitat is shrinking as
human populations expand. However, wild
dogs are able to survive in places that are of
marginal use to man, such as semidesert
and swamps, and they breed successfully in
zoos.   JM

A Hunting Tradition

A zebra stallion trots away from his group of mares and foals as the pack of dogs approaches. With head lowered, teeth bared and nostrils flaring he charges at the leading dogs, who turn and flee.

This is usually what happens when African

wild dogs and zebras meet. But not always. A minority of wild dog packs will attack zebras. When they do, it is they who charge first, to cause a stampede before the stallions can take the initiative. Then they mount a closely coordinated attack (see drawing), one dog grabbing the chosen victim's tail and another its upper lip, while the rest disembowel it.

Of 10 wild dog packs recently studied on the Serengeti Plain in Tanzania, only two had the ability to turn the tables on an adult zebra—eight times a dog's weight—and kill it. These two zebra-hunting packs were large, with eight or more adult members. but pack size

was not the crucial factor: three other packs of similar size ignored zebras and, on one occasion, just four dogs from a zebra-hunting pack were observed to kill a zebra.

Why then do not all African wild dog packs hunt zebras? A clue to the answer seems to lie in the fact that one of the two zebra-hunting packs was known to have hunted zebras for at least 10 years, over three generations. For this pack, zebra-hunting was a tradition, learned by each generation from its predecessor. Other packs studied did not exhibit any such hunting tradition, although some did show a preference for Grant's gazelles over Thomson's gazelles.

Not only zebra hunting but also such knowledge as the location of water, prey concentrations and range boundaries may be passed on as a tradition. Studies of African wild dogs (and also of some monkeys) indicate that man's previously supposed unique reliance on cultural as against genetic inheritance should rather be viewed as a dramatic extension of a pattern that exists in other social animals.

▲ **The chase.** A pack of African wild dogs hard on the heels of a wildebeest, which has probably been selected because it is young or weak. The two leading dogs, about to seize the prey, are almost certainly the dominant individuals within the pack, as these take the most dangerous role in the hunt.

▼ **The kill.** Once caught, prey is rapidly disemboweled – here the fate of a Thomson's gazelle.

A "Back-to-front" Social System

Parent role-reversal in African wild dogs

In a strange tug-of-war, one adult female African wild dog holds a puppy by the head, another has the pup's hind-quarters in her mouth, and the two are pulling in opposite directions. Pups are often injured and may even die in these struggles, which are part of a protracted conflict over possession of pups when two females try to raise litters at the same time. Such macabre behavior is just one example of the strange social organization of this species, in which only one female per group normally breeds successfully.

The social arrangements of African wild dogs are extraordinary because they are the exact opposite of those in most other social mammals, such as coatis, baboons, lions and elephants. Within the wild dog pack all the males are related to each other, and all of the females to each other but not to the males. Females migrate into the pack, whereas males stay with their natal pack. Before entering a pack, female littermates live in friendly coexistence, but once they start to compete to breed aggression breaks out. Males in the pack outnumber females by two to one—more males are born (59 percent of young born in one study) than females, and aggression between females often leads to the disappearance, and probably the death, of the loser. Only the dominant male and female normally breed, but should another female produce a litter few if any will survive because of harassment by the dominant female.

Why have African wild dogs evolved such a "back-to-front" social system? The evolutionary success of any species and of its individual members can be measured by the number of young successfully reared. However, the number that can be reared is limited by the amount of effort that can be expended by parents in raising young—the parental investment. Each sex must maximize its parental investment if it is to be reproductively successful. In most social mammals the female has a major, often exclusive, role in rearing offspring, and her ability to provide adequate food and protection is limited to a few young. In such systems the males provide only a single sperm cell; so to achieve maximum reproductive success males mate with many females. The capacity of parental investment of the females is the limiting resource, over which the males fight. The overall result is that the maximum number of young is reared within the limits of parental investment.

The African wild dog achieves the same results, but with a reversal of roles. Since

males help raise young and there are more of them than females, the number of offspring reared is controlled by their parental investment, not the females'—this time males are the limiting resource and it is the females that fight to ensure that it is their young that the males help to rear, not another female's.

The question can now be asked, How did this system come about? It seems likely that African wild dogs evolved from a jackal-like ancestor which showed some degree of parental care (as do jackals today—see p58). Later, African wild dogs became specialist cooperative hunters of large hoofed animals, so that individuals could only survive in a pack. However, to prevent the adverse effects of inbreeding, one sex has to disperse. In this case it is the females that do so; but

▲ ▶ **Aggression breaks out** when females compete to breed. Simple gestures (as with the female threatening a throat bite to her sister BELOW RIGHT) help establish which female should breed. If the dominance hierarchy breaks down, and a second female breeds, the dominant female will fight over pups of the other litter ABOVE, which often results in death for the pups.

Comparison of breeding systems

African wild dogs and lions are cooperative hunters which inhabit the same regions of Africa and rear young communally. The pattern seen in lions is typical of most social mammals.

Lion	African wild dog
Stable groups composed of related females: daughters stay with mothers, aunts etc.	Stable groups composed of related males: sons stay with fathers, uncles etc.
Males usually leave natal group and try to breed elsewhere.	Females usually leave natal group and try to breed elsewhere.
Aggression more intense between males than females.	Aggression more intense between females than males.
More females survive to maturity than males.	More males survive to maturity than females.

◄ ▼ **Friendly relations prevail** in a pack most of the time, as in the ritualized so-called midday greeting ceremony. All members of the pack run around excitedly, squeaking and thrusting their muzzles into each other's faces LEFT, a gesture which derives from infantile begging (detail, BELOW). At other times adult male pack-mates and juveniles join in friendly playful fights BOTTOM RIGHT.

why not males? Even though all adults help to rear the young, the pack still only has the capacity to raise one litter. If more are produced the pack will be over-extended and few if any pups survive. Thus in such circumstances only one female has reproductive potential; the only chance for a second female to produce young is to migrate to another pack, where she might achieve dominance.

It is interesting that competition between females has also been reported to be more severe than between males in wolves and coyotes, both species with considerable male investment in pup-rearing and anatomically similar to the ancestor of the African wild dog. The prolonged conflicts seen today between two would-be breeding African wild dog females may be a vestige of the original agent of natural selection that first led this species to evolve its peculiar breeding system.   JM

DHOLE

BRANDED as cruel and wanton killers because they often kill by disemboweling their prey, dholes are persecuted by man throughout their range. Until recently, little was known of these secretive animals, but it has now been established that they are group-living, with cooperative hunting and group care of young at the heart of their societies. In many respects their life-style resembles that of the African wild dog (see pp68–71).

The dhole differs from most canids in having one fewer molar teeth on each side of the lower jaw and in having a thick-set muzzle, both adaptations to an almost wholly carnivorous diet. They feed on wild berries, insects, lizards and on mammals ranging in size from rodents to deer. They eat fast: a fawn is dismembered within seconds of the kill. At a kill, dholes compete for food chiefly through speed of eating, rather than by fighting. A dhole can eat 8.8lb (4kg) of meat in 60 minutes, and it is common to see dholes running from the carcass with pieces of meat to eat undisturbed by other pack members. Heart, liver, rump, eyeballs and any fetus are eaten first. When water is nearby, dholes drink frequently as they eat. If water is distant, they make for it soon after eating. Dholes will often lie in water even in the cool of the day. They do not cache food, but they often scavenge their own kills and those of leopards and tigers.

A dhole pack is an extended family unit, usually of 5–12 animals, and rarely exceeding 20. In a pack observed in Bandipur Tiger Reserve, southern India, the average number of adults was 8, rising to 16 when there were pups; there were consistently more males than females. Packs are territorial and numbers are regulated by social factors affecting reproduction (only one female breeds) and by emigration or deaths in both adults and young.

Dholes are sexually mature at about one year. Whelping occurs between November and April and the average litter size is eight. Before giving birth the bitch prepares a den, usually in an existing hole or shelter on the banks of a streambed or among rocks. In the Bandipur pack more than three adults took

Hunting Strategies

Dholes are active chiefly during the day, although hunts on moonlit nights are not uncommon. Most hunts involve all adult members of a pack, but solitary dholes often kill small mammals such as a chital fawn or Indian hare. Prey is often located by smell. If

tall grass conceals their prey, dholes will sometimes jump high in the air or stand briefly on their hind legs in order to spot it.

Dholes have evolved two strategies to overcome the problems posed by hunting in thick cover, both depending heavily on cooperation in the pack. In strategy 1, the pack moves through scrub in extended line abreast, and any adult capable of killing when it locates suitable prey may begin the attack. If the prey is small it will be dispatched by one dhole. When the prey is larger—for example a chital stag—the sound of the chase and the scream of the prey attract other pack members to assist. It is rare for two large animals to be killed in one hunt in the scrub.

In strategy 2 (LEFT) some dholes remain on the edge of dense cover to intercept fleeing prey as it is flushed out by the other pack members. In thick jungle the chase seldom lasts more than 0.3mi (0.5km).

Larger mammals are attacked from behind, usually on the rump and flank, and immediately disemboweled. The resultant severe shock and loss of blood kills the prey—dholes seldom use a throat bite. Small mammals are caught by any part of the body and killed with a single head shake by the dhole. Even before their prey is quite dead, dholes start eating. In general they are efficient killers—two or three dholes can kill a deer of 110lb (50kg) within two minutes. Interference at this stage by human observers will prolong the death throes, thus fueling the prejudice that dholes are cruel hunters.

▲ **Dholes feeding on a chital stag.** Pack members normally feed peaceably at kills, but individuals may sometimes, as here, remove portions to consume undisturbed at a distance.

▼ **Before the morning hunt** a pack of adult dholes, calling intermittently to one another, gather in a forest clearing. The hunting strategies of dholes depend on close cooperation between pack members (see box).

part in feeding both the lactating mother and the pups, which eat regurgitated meat from the age of three weeks. At this time the hunting range of the Bandipur pack was some 4sq mi (11sq km), much smaller than its normal range of 15sq mi (40sq km). Sometimes a second adult (the so-called "guard dhole") stays by the den with the mother while the rest of the pack is away hunting.

Pups leave the den at 70–80 days; if the site is disturbed earlier the pups will be moved to another den. The pack continues to care for the pups by regurgitating meat, providing escorts, and allowing pups priority of access at kills. At five months pups actively follow the pack and at eight months they participate in kills of even large prey such as sambar, a species of large deer.

Dhole calls include whines, growls, growl-barks, screams and whistles, and squeaks by the pups. The whistle is a contact call most often used to reassemble the pack after an unsuccessful hunt. The "latrine sites"—communal defecation sites at the intersection of trails and roads—may be a major means of communication by smell, serving to warn off neighboring packs at the edge of the home range and to mark how recently an area has been hunted, thus ensuring efficient use of all parts of the home range.

Hunters consider dholes as rivals, jungle tribesmen pirate their kills, and until recently others tried to eliminate them by poisoning their kills and offering bounties. Today the main threat to the dhole comes from habitat destruction and decimation of prey species by man. In India the creation of many tiger reserves and national parks has helped to conserve the subspecies *C. a. dukhunensis.* AJTJ

MANED WOLF

Chrysocyon brachyurus [v]
Maned wolf, *lobo de crin*, or *lobo guará* or *boroche*.
Sole member of genus.
Family: Canidae.
Distribution: C and S Brazil, Paraguay, N Argentina, E Bolivia, SE Peru.

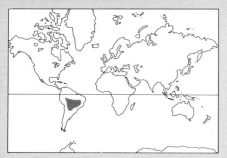

Habitat: grassland and scrub forests.

Size: head-body length 41in (105cm); tail length 18in (45cm); shoulder height 34in (87cm); weight 51lb (23kg). No variation between sexes or different populations.

Coat: buff-red, with black "stockings," muzzle and "mane"; white under chin, inside ears and tail tip. Pups born black but with white-tipped tail.

Gestation: about 65 days.

Longevity: unknown in wild (12–15 years in captivity).

[v] Vulnerable.

IN Brazil, the cry of the Maned wolf at night is believed to portend changes in the weather and its gaze is said to be able to fell a chicken. These are two of the myths that shroud South America's largest and most distinctive canid. Although considered endangered throughout its range, it remains one of the least studied of wild dogs.

The Maned wolf is so named for the patch of long black erectile hairs across the shoulders and for its wolf-like size. But it is not a true wolf (see p50), and most closely resembles in general form and coloring a long-legged Red fox. The tail is relatively short, the ears are erect and about 7in (17cm) in length, and the coat is softer in texture than that of many canids and lacks underfur.

It has been suggested that its long legs are an adaptation for fast running. In fact, Maned wolves, which have a characteristic loping gait, are not particularly swift runners and their long legs are most likely an adaptation to tall grassland habitats. The foot pads are black and the two middle toe pads are joined at the base. To increase the area of contact with marshy ground, the foot can be spread laterally.

Maned wolves are opportunists, taking small vertebrate prey up to the size of pacas, which weigh about 18lb (8kg). Rabbits, small rodents, armadillos and birds are the most common prey, with occasional fish, insects and reptiles. Seasonally available fruits make up about half the diet, the most frequent being *Solanum lycocarpum*, known as "fruta do lobo"—wolf's fruit, which may have therapeutic properties against the Giant kidney worm (*Dioctophyma renale*) common in the Maned wolf. Foraging is usually done at night, but sometimes occurs in the day in areas less disturbed by man. Individuals hunt alone and may cover 20mi (32km) during the course of a night. They catch small vertebrates by using a slow stalk followed by a stiff-legged pounce similar to that of Red foxes.

Although sexually mature after about one year, Maned wolves probably do not breed until nearly two years old. Females produce one litter (2–5 pups) per year, usually in June–September. They reach adult size by about one year. Maned wolves make their dens in available cover, for example, tall grass or a thicket. The extent to which males take part in raising the young is not known for free-living individuals but males in captivity have been observed to care for pups and feed them by regurgitation. Females probably also regurgitate food to young. The breeding of this species in captivity has rarely been successful.

Little is known about the social organization of free-living Maned wolves. However, one study suggests that two adjacent, but nonoverlapping, territories of about 11sq mi (30sq km) were each occupied by a monogamous pair. Although the male and female of each pair shared the same range, they were rarely found in close association except during the breeding season. In captivity, serious fighting often occurs when individuals of the same sex are placed in the same enclosure. Maned wolves in the wild have been seen to deposit feces at intervals along major pathways and to renew this marking periodically. Special defecation sites are used near favorite resting places

and dens, where feces are deposited at an average height of 16in (40cm) above surrounding ground level.

The Maned wolf is classified by the IUCN as "vulnerable" and by the Brazilian government as "endangered." The species' range has diminished considerably during recent decades. Maintaining these animals in captivity is complicated by the need for enclosures of at least several hundred square meters. In addition, Maned wolves are subject to a variety of diseases including parvovirus as well as Giant kidney worm infestation. About 80 percent of captive and sampled wild-caught individuals suffer from cystinuria, an inherited metabolic disease known to be fatal in some cases.

Maned wolves are only occasionally hunted for sport, but are frequently captured for sale to South American zoos. Individuals may take farm stock up to the size of a lamb and the species is shot or trapped as a result. Although these canids are shy and usually avoid man, female Maned wolves have been known to defend pups aggressively against capture by humans. In Brazil, parts of the Maned wolf's body, even its feces, are supposed to have medicinal value, or to work as charms. For example, the left eye removed from a live Maned wolf is said to bring luck.   JMD

▲ **Unexpectedly long legs,** attractive red coat and rarity value of the Maned wolf make it a popular exhibit at zoological gardens fortunate enough to have one.

◀ **Sizing each other up,** two Maned wolves, here in captivity, circle each other warily. The animal on the right, with its arched back, erect hair and turned head is giving an impression of greater size, probably in an effort to outface the other.

◀ **Surveying its native wilderness,** a Maned wolf displays itself on the skyline. Maned wolf numbers are dwindling and the species is now regarded as threatened.

Evolution of the Fox-on-stilts

A poor fossil record has obscured the origin of the Maned wolf. However, it has been suggested that one or more waves of small primitive canids may have invaded South America from North America some 2 million years ago, and that these early savanna canids were probably faced with two locomotion options. They would have to go either through the tall grass, or over the top of it. Apparently only the Maned wolf took the latter route. With long legs and height came large body size and a heavy additional energy requirement. Opportunistic foraging, large mutually exclusive territories, and a tendency to pair with a single mate may all be adaptations to a scarce and evenly distributed food supply. In a tropical climate, a large body also brings with it the problems of temperature control. The pups' black fur, lack of underfur in adults and the nocturnal activity of the Maned wolf may all have evolved in response to this problem.

As in other large carnivores of the grasslands, such as the lion, the Maned wolf has evolved simple methods of long-distance communication. When wanting to be conspicuous, an individual turns broadside, erecting the hair across its shoulders and along its back (see ABOVE); the back is arched and the head turned to the side to display the white patches on the ears and throat. The cry of the Maned wolf is a deep-throated extended bark repeated at intervals of about seven seconds. These signals may have evolved in part as aids to maintaining a certain level of dispersal among individuals of the species.

OTHER DOGS

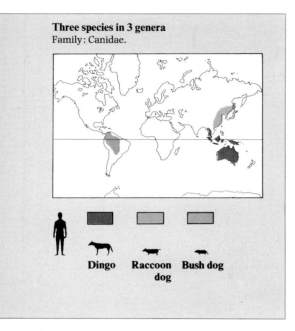

Three species in 3 genera
Family: Canidae.

Dingo   Raccoon   Bush dog
          dog

▼ **Wild dog of Australia. The dingo** is descended, like the Domestic dog, from the wolf and lives as a completely wild dog in Australia; similar forms inhabit many Southeast Asian islands.

THE **dingo** may be descended from the earliest tamed members of the genus *Canis*. Its origins remain obscure although it has lived unchanged in Australia for at least 8,000 years. The dingo's history has always been loosely associated with the Aborigines who probably colonized Australia some 20,000 years before it did. Perhaps Australian Aborigines used the dingo for warmth at night or as food or a guard, since they probably did not use it for hunting. The flexibility of dingo social behavior parallels that of the coyote or wolf (see pp50–53); indeed, the dingo is probably descended from the Indian wolf *C. lupus pallipes*. Its diet varies between small mammals, especially European rabbits, lizards and grasshoppers caught by one individual to large ungulates, especially feral pigs and kangaroos, devoured as carrion or hunted by a pack.

Although the name dingo is most commonly applied to the wild canid of Australia, we use it to describe also the suite of little-known "dingo-like" dogs, members of which have been variously called the New Guinea singing dog, Malaysian wild dog, Siamese wild dog, Filipino wild dog etc. The term pariah dog is sometimes loosely used to cover both wild dingoes and feral domestic dogs. Some taxonomists argue that all dingoes are feral domestic dogs (*C. familiaris*). Here we opt for the other view, that dingoes are sufficiently distinct from domestic dogs, anatomically and reproductively, to merit species status.

The New Guinea singing dog, found in mountains above 7,000ft (2,130m), is sometimes known as *C. hallstromi*, and may be descended from the ancient Tengger dog of Java (a fossil ancestor of dingoes). It howls rather than barks, and is used by Papuan natives as food, for hunting and as a guard.

There is a risk that these small populations of "dingo" will be outbred to oblivion with village curs before they can be studied. This is sad because not only are they interesting in their own right, but they are important for unraveling the history of domestication of the dog and of primitive man's migrations.

In Australia the dingo is economically important, directly because of attacks on sheep and thus indirectly due to the costs of poisoning, bounties and dingo-proof fencing. Aerial poisoning with baits has greatly reduced dingo numbers, but their populations are generally robust in Australia, except in Queensland where remnants of "pure dingo" are likely to be swamped to extinction by feral dog genes.

Raccoon dogs, as their name suggests, look rather like raccoons (family Procyonidae). Although the natural distributions of the two species are widely separated, in Europe confusion can occur since both species have been introduced there. Some taxonomists regard Raccoon dogs as primitive canids, but their affinities with other living species are unknown.

Raccoon dogs are omnivores, with a diet which varies with the seasonal availability of fruits, insects and other invertebrates, occasional vertebrates and, where available, marine invertebrates caught while beach-combing. Apparently they prefer to forage in woodland with an abundant understory, especially of ferns. In the 1920s they were introduced into western Russia for fur farming (furriers call them Ussuri raccoons); they subsequently spread throughout much of eastern Europe and north to Finland. Outside the fur trade this range expansion has been viewed as undesirable, since the Raccoon dog carries rabies and its arrival has further complicated the control of Red fox rabies in eastern Europe. There is some evidence that during their winter hibernation—itself a unique feature amongst canids—Raccoon dogs may incubate the rabies virus and hence cause the disease to

▲ **Barely resembling a dog, the Bush dog** of Central and South America has short legs, a compact body, short snout and small ears.

▲ **Dingos live in packs,** like coyotes or wolves, and may hunt cooperatively.

▼ **Looking more like a raccoon than a dog,** the Raccoon dog of Eastern Asia has been introduced to parts of Europe.

persist from one season to the next, in places where fox densities are so low that rabies might otherwise die out.

Although they are the only members of the family reputed not to bark, the behavior of Raccoon dogs is otherwise recognizably canid. Like Bat-eared foxes, they occasionally hold their tails aloft in inverted-U positions during social interactions. Subordinate animals apparently do not wag or lash their tails, unlike other canids. However, following mating there is a copulatory tie.

Raccoon dogs are reported to live in pairs or temporary family groups, with males contributing to pup care. Preliminary radio-tracking studies suggest that the home ranges of several Raccoon dogs may overlap widely and when food is clumped, such as at a fruiting bush, they may feed together amicably. In two study areas in Japan home ranges averaged about 25 and 123 acres (10 and nearly 50 hectares), whereas one estimate from Europe is of 250–500 acres (100–200 hectares). All these figures indicate relatively small ranges, considering the species' body size. Raccoon dogs defecate

at latrine sites, each dog using several, but not all, of the latrines within its home range. Animals feeding at the same place do not necessarily use the same latrines. These latrines are clearly a form of scent marking. The highest rates of marking occur in summer. Although the mated pair is thought to be the basic social unit, there is no firm evidence on the ties between animals that share all or part of their home ranges.

The **Bush dog** is one of the least known of canids and one of the least "dog-like" in appearance. It is a stocky, broad-faced animal with small, rounded ears, squat legs and a short tail.

Bush dogs are elusive and very little is known of their behavior in the wild. They are, however, listed as vulnerable by the IUCN.

Packs of up to 10 animals have been reputed to hunt together, often seeking prey considerably larger than themselves, for example capybaras and rheas (to 100lb/45kg and 55lb/25kg respectively). Packs are said to pursue amphibious prey into the water, where the dogs swim and dive with agility. Their teeth are sturdy and highly adapted to their carnivorous way of life. The dental formula (I3/3, C1/1, P4/4, M1/2 = 38) is not shared by any other American canid.

Studies in captivity have shown that males feed their nursing mates and that littermates squabble very little over food in comparison with, say, young foxes. This is probably related to their cooperative hunting in later life—wolf pups are similarly amicable over food. Adults are said to keep in contact with frequent whines; this may be an adaptation favorable to maintaining group cohesion whilst foraging in forest undergrowth where visibility is poor. When scent marking, males urinate by cocking their legs at 90 degrees, but females reverse up to trees and urinate on the trunk from a handstand position.   DWM

Abbreviations: HBL = head-body length: TL = tail length: wt = weight. [v] Vulnerable.

Dingo
Canis dingo

Australasia, including Indonesia (part), also Malaysia, Thailand, Burma. Ubiquitous, from tropical forest to semiarid regions. HBL 59in; shoulder height 20in; TL 14in; wt 44lb. Coat: largely reddish-brown with irregular white markings. Gestation: 63 days. Longevity: up to 14 years in captivity; widely subject to bounty schemes in wild.

Raccoon dog
Nyctereutes procyonoides

E Asia, in Far East, E Siberia, Manchuria, China, Japan and N Indochinese Peninsula. Introduced in Europe. In woodland and forested river valleys. HBL 20–24in; shoulder height 8in; TL 7in; wt up to 16lb. Coat: long brindled black-brown body fur with black facial mask, sleek black legs and black stripe on tail. Gestation: 60–63 days. Longevity: not known. Subspecies: 5.

Bush dog [v]
Speothos venaticus
Bush dog or Vinegar fox.

Panama to Guiana and throughout Brazil. In forests and forest-edge marshland. HBL 26in; shoulder height 10in; TL 5in; wt 11–15lb. Coat: dark brown; lighter fawn on head and nape; chin and underside may be cream-colored or dark. Gestation: reportedly 80 days or more. Longevity: not known. Subspecies: 3.

THE BEAR FAMILY

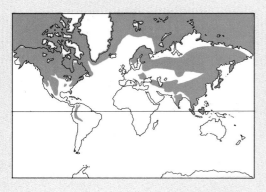

Family: Ursidae
Seven species in 5 genera.
Distribution: Arctic, N America, Europe, Asia and S America.

Habitat: from Arctic coasts to tropical jungle, chiefly forests.

Size: ranges from 3.6–4.6ft (1.1–1.4m) in overall length and weighing 60–143lb (27–65kg) in the Sun bear to 6.6–9.8ft (2–3m) and 220–1,430lb (150–650kg) in the Polar bear, fat-laden males reaching 1,760lb (800kg) or more.

Grizzly or **Brown bear** *Ursus arctos*
Polar bear *Ursus maritimus*
American black bear *Ursus americanus*

Asian black bear *Selenarctos thibetanus*

Sun bear *Helarctos malayanus*

Sloth bear *Melursus ursinus*

Spectacled bear *Tremarctos ornatus*

THE bears (family Ursidae) include the world's largest terrestrial carnivores—the Grizzly and Polar bears. Most bears are found in small concentrations in temperate and subtropical regions; they are seldom numerous, but their size and activities command man's attention.

The four genera that contain single species are more southerly in distribution than the three *Ursus* species. The Sun bear is the smallest, with the Sloth bear, Asian black bear and Spectacled bear similar in size. *Ursus* species increase in size from the American black bear through the Grizzly bear to the Polar bear, males of which may weigh over 1,760lb (800kg). In species reported to have only one mate (Sun and Sloth bears), males are only 10–25 percent larger than females, but in species whose males may compete for access to several females they are 20–100 percent heavier.

Bears have large, heavily built bodies, thick, short, powerful limbs and short tails (rarely over 4.7in/12cm long). Their eyes and erect rounded ears seem small in comparison to their large heads. Hearing and sight are less developed than their acute sense of smell. Bears walk on the soles of their feet (plantigrade gait) which are broad, flat and armed with five long, curved nonretractile claws, used while foraging or for climbing. Apart from the Polar bear and grizzly, most species climb well, but the Sun bear is the most adept, helped by its naked soles. In other bears the soles are well furred, especially in the Polar bear. Bears can run fast over a short distance. The coat is long, shaggy and predominantly one color—black, some shade of brown, or white. The four smaller species have lighter chest markings.

Bears are largely herbivorous, as their build and the arrangement of their teeth suggests, although they may take meat occasionally. The one exception is the Polar bear, which feeds primarily on seals. Sloth and Sun bears apart, bears generally forage by day. Unlike canids, bears use few vocalizations in communication and few facial expressions. (It is a matter of controversy whether pricked or lowered ears presage a ·change.) Scent marking is known to be used by, for example, the grizzly (see box, p82).

Births relate to seasons in temperate and arctic regions but not in the tropics. Most species are promiscuous, with male home ranges overlapping many female home ranges. True gestation is short, although implantation of the fertilized egg is delayed, so that apparent gestation is long, up to 265 days in the Polar bear. The young (usually 1–3) are born in seclusion, near-naked (except the Polar bear), helpless and very small (7–25oz/200–700g).

Bears from colder regions (*Ursus* species and some Asian black bears) enter a period of winter dormancy or lethargy, during which time the cubs are born. This behavior is not true hibernation as body temperature and pulse rate do not drop. However, the bears do not eat during this time but live off fat built up during a period of enormous appetite in the fall. The reasons for this winter lethargy are two-fold. Firstly, the bears' chief food (succulent vegetation) is not available in the cold northern winters. Secondly, the cubs are so small at birth (0.25–1 percent of the mother's weight) that they cannot regulate their own body temperature; the snug environment of the den provides the warmth necessary to prevent death. The exceptions to "hibernation" highlight these reasons; among Polar bears (which eat meat and therefore do not rely on seasonal vegetation) only pregnant females den, and in southern populations of Asian and American black bears, where food is available throughout the year, the males remain active in winter.

Bears evolved more recently than other carnivores. They can be traced back 20 million years to the Miocene era when the Dawn bear (*Ursavus elmensis*) first appeared—a small carnivore about the size of a

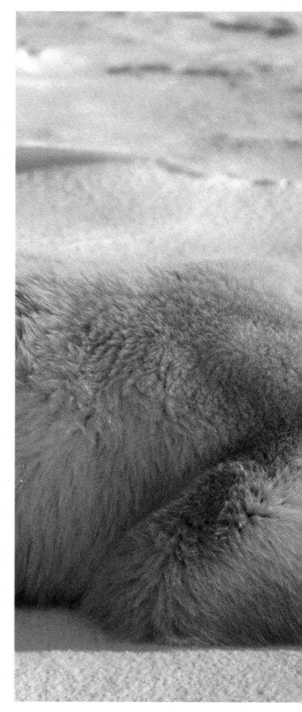

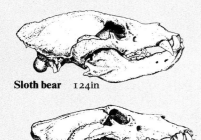

Sloth bear 124in

Grizzly 146in

Skulls of Bears

The skulls of bears are the longest and most massive among the carnivores. Bears' teeth show how the family has evolved from being chiefly carnivorous to largely herbivorous animals. The Grizzly bear has the typical bear arrangement of unspecialized incisors, long canines, reduced or absent premolars (the carnassial teeth undeveloped), and broad, flat molars with rounded cusps for crushing the vegetable matter that makes up much of the diet. The wide variation in number of teeth between individuals (I3/3, C1/1, P2–4/2–4, M2/3 = 34–42) indicates that a reduction in the number of cheek teeth is in mid-evolution.

However, the Polar bear, which is the most recent of the *Ursus* line, has a purely carnivorous diet (chiefly seal meat) and appears to be evolving flesh-shearing carnassials again.

The Sloth bear has an unusual dental formula for a bear. It lacks the inner pair of upper incisors (I2/3)—a specialized modification for sucking up termites. The smaller molars are an adaptation to a diet of fruit and insects.

fox. Gradually, in later species, the skull became more massive, the carnassial teeth lost their shearing function, the molars became square with rounded cusps, weight increased, the tail was reduced to a mere stub, legs and feet became heavy, the feet relatively shorter, and the gait plantigrade. *Ursus minimus* appeared about 5 million years ago and probably gave rise to the Asian and American black bears. Brown bears first appeared in Asia and then spread to Europe some 250,000 years ago, where they confronted the Cave bear (*U. spelaeus*). Later (70,000–40,000 years ago) they reached North America. Before then, however, the grizzly gave rise to the Polar bear, the most recent of the *Ursus* line, with no fossil record before 70,000 years ago.   FB

▲ **Grizzly bears at a salmon run.** The bears only congregate in numbers at such sites of food concentration.

◄ **Growing to be the largest of the bears,** Polar bear cubs are only 0.25 percent of the weight of their mother at birth.

GRIZZLY BEAR

Ursus arctos

Grizzly or Brown bear.
One of 3 *Ursus* species.
Family: Ursidae.
Distribution: northwest N America;
Scandinavia to E USSR; Syria to Himalayas;
Pyrenees, Alps, Abruzzi and Carpathian
mountains.

Habitat: chiefly forests.

Size: variable, depending on locality and nutrition; adult females with access to salmon weigh 265–453lb (120–205kg), inland females 176–397lb (80–180kg). Males 20–80 percent heavier, largest recorded from British Columbia and Alaska 850 and 975lb (386 and 443kg); head to tail-tip length 6.6–9.2ft (2.0–2.8m); shoulder height 4–5ft (1.2–1.5m).

Coat: long, coarse, usually brown and frequently white-tipped (grizzled) but color variable from cream to black.

Gestation: 210–255 days.

Longevity: to 25–30 years in wild (47 recorded in captivity), but usually less.

Subspecies: 2 or 3. **Grizzly bear** (*U. a. horribilis*) or **Brown** or **Eurasian brown bear**—males rarely over 600lb (275kg)—over whole of range except that of Kodiak bear. **Kodiak bear** (*U. a. middendorffi*) restricted to Kodiak, Afognak and Shuyak islands. **Eurasian brown bear** sometimes considered separate subspecies (*U. a. arctos*).

GRIZZLIES may weigh nearly half a tonne, can shear off half-inch steel bolts and can charge at 32 miles an hour; but these largest of land animals are also one of the most vulnerable species on earth.

The Grizzly bear (or Brown bear as it is mostly known in Europe) once ranged throughout North America, most of Europe, and northern Asia. Today, although it remains the most widely distributed bear, large populations remain only in Alaska and Canada and in the USSR. Today's population of 1,000 or more in the southern USA has declined from perhaps 100,000 in 180 years. There are some isolated populations in Europe, amounting to fewer than 30 individuals.

In America the largest individuals are in the far Northwest (Kodiak bears). In Eurasia size increases from west to east (Kamchatkan brown bear). But size is notably related to available food, and subspecies distinctions also appear to be more related to nutrition than to geographic isolation.

► **Lord of his domain,** this massive Grizzly bear need fear no other animal apart from man. When surveying their surroundings or out to intimidate, grizzlies often assume such an upright, totem-like posture; otherwise they travel on all fours.

▲ **The size and power** of the Grizzly bear are fully expressed in the massive proportions of this bear's head. The prominent nose and diminutive ears reflect a reliance on smell rather than hearing.

▼ **Grizzlies love water**: Here a satiated adult relaxes in the shallows of a salmon stream.

The fur is extremely variable in color, from cream through cinnamon and brown to black. In gross form the Grizzly bear has a concave outline to the head and snout, ears that are inconspicuous on a massive head, and high shoulders which produce a sloping backline. Its sense of smell is much more acute than its hearing or sight.

During the 4–7½ months spent outside their den (much more in southern populations) grizzlies consume large amounts of food—26–35lb (12–16kg) a day. Grizzlies cannot digest fibrous vegetation well and they are highly selective feeders. The diet shows dramatic shifts as they move from alpine meadows to salmon streams to avalanche chutes and riverside brushlands. Grizzlies are omnivorous, with flattened cheek teeth and piercing canines 1.2in (30mm) or more in length (see p87). Their large claws often exceed 2.4in (6cm) in length; they are used to dig up tubers and burrowing rodents. The diet is dominated by vegetation, primarily succulent herbage,

tubers and berries. Insect grubs, small rodents, salmon, trout, carrion, young hoofed mammals (deer etc) and livestock are all taken as the opportunity arises.

Breeding occurs in May or June, when males search for receptive females. Ovulation is induced by mating, after a brief courtship of 2–15 days. Implantation of the fertilized egg is delayed until October or November, when the female dens in a self-made or natural cave, in a hollow tree or under a windfall. The young (usually 2–3) are born in January–March; they weigh only 12.5–14oz (350–400g), are nearly naked, and are quite helpless. They remain denned until April–June, then accompany the mother for 1½–4½ years. The age at which a female first gives birth, the litter size and the interval between litters are controlled by nutritional factors (see box, p87). Population numbers are largely independent of population density among females but density-dependent among males. Because reproduction is under nutritional

▲ **Adept salmon fishers,** Grizzly bears catch their prey with teeth or claws and usually take it ashore before delicately stripping off the flesh, first on one side, then the other, leaving behind the head, bones and tail.

▶ **The bear hug.** Apparently locked in vicious combat, two juvenile males practice their fighting skills. Later in life such fights are for real, possibly with a fatal outcome.

control, females tend to establish exclusive access to forage and mutually exclusive home ranges, although limited overlap may occur. The range sizes of adult females can vary enormously from region to region; for example, 5.5sq mi (14.3sq km) on an Alaskan island and 73sq mi (189sq km) in northern Alberta. Comparable figures for adult males are 9.4sq mi (24.4sq km) to 406sq mi (1,054sq km). Young females may remain in the mother's range after leaving her care—three generations of females have been known in the same range. Adult males are solitary, and their ranges encompass those of several adult females, as well as overlapping with other adult males. Young males may travel up to 62mi (100km) after leaving their mothers and spend much of their time avoiding risky contact with adult males.

The Grizzly bear has been wiped out over much of its range and is endangered in many areas. Hunting and loss of habitat are the major causes. Legal hunting can be controlled, but bears that venture into man's expanding domain are killed because of the threat (actual or feared) to livestock. The use of incinerators and bear-proof disposal units in parks, rather than dumps, can reduce available food. Townsites and livestock are simply incompatible with bears.

FB

Why Bears Are Aggressive

Every year big bears maul or kill other bears—and sometimes people. Bears *are* aggressive. Why? The reason, true for both males and females, lies in the drive to achieve reproductive success. Female Grizzly bears in northern inland populations are unlikely to produce more than 6–8 young during their lives; and even the healthiest American black bear female is unlikely to produce more than 12–13. Thus, each young bear is crucial to a female's reproductive success and is vigorously defended. Other bears or people who stray close enough to appear as a threat to young bears are charged and may be wounded or killed.

Males fight to ensure they sire as many cubs as possible, and thus perpetuate their own genes. As receptive females are rare, scattered, and only available for breeding every few years, at which time they are promiscuous and likely to conceive young sired by different males, the male has to decide whether to defend a single female from the attention of other males or to mate with as many females as possible. Bears take the latter option. Defending a female from all comers would bring a male into conflict with other equally mature males—something to be avoided, as such fights, when they do occur, often result

in the death or severe wounding of one combatant.

Males reduce potential competition by evicting from their home ranges (or even killing) subadult males that might compete for females in later seasons. Out of the breeding season, mature males also establish a dominance hierarchy of access to females; this occurs particularly when mature males gather at sites of food concentration—for example, at waste dumps or salmon runs.

Marking probably serves to advertise the presence of a dominant male and thus to reduce the risk of dangerous encounters. In areas of stable air currents, bears mark by scraping the bark off trees and rubbing against the surface, leaving scent. Where air currents are unstable, bears often mark the ground with regularly spaced depressions or scrapes.

Bears' eyesight is poor; a myopic bear may not distinguish between humans and subadult bears so most attacks on man are probably cases of mistaken identity. Although human body odors are generally repellent to them, bears living near waste dumps without incinerators generally associate human odor with food—often with tragic consequences. Such human victims are rarely eaten.

POLAR BEAR

Ursus maritimus [v]
One of 3 *Ursus* species.
Distribution: circumpolar in northern
hemisphere.

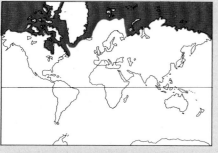

Habitat: sea ice and waters, islands and coasts.

Size: body length of males
8.2–9.8ft (2.5–3m),
females 6.6–8.2ft
(2–2.5m). Weight of males
770–1,430lb
(350–650kg) or more in
fat-laden individuals,
females 385–660lb
(175–300kg).

Coat: white, or yellowish from staining and
oxidation of seal oil.

Gestation: about 8 months.

Longevity: 20–25 years.

[v] Vulnerable.

▶ **Four-square, the largest carnivore on land,**
a Polar bear is captured in statuesque pose.
Atop its back is an immobilizing dart, fired in
order to tag the bear for conservation purposes.

▼ **Arctic sun silhouettes four bears.** Usually
solitary, Polar bears may congregate in ice-free
conditions or by major food sources.

THE Polar bear inhabits a cold and hostile
environment that most of us never see.
Yet Polar bears, part of the culture of arctic
coastal peoples, are of special interest to
increasing numbers worldwide who wish to
see this species preserved.

Not only is the Polar bear the largest
carnivorous quadruped, it is also unique in
its combination of great size, white color and
adaptations to an aquatic way of life. Polar
bears are as big as or bigger than the large
Brown bears, but less robustly built, with a
more elongated head and neck, and are
adapted to a sea ice environment. A thick
winter coat and fat layer protect them
against cold air and water. They are
completely furred except for the nose and
the foot pads; short ears are another adaptation to cold. Polar bear milk has a high fat
content (31 percent) and enables cubs to
maintain their body temperature and to
grow rapidly during the four months before
they leave the den. The white coat color
serves as camouflage and the claws are
extremely sharp, providing a secure grasp
on the bears' seal prey. Their acute sense of
smell is an essential aid to hunting. Polar
bears can, if necessary, swim steadily for
many hours to get from one piece of ice to
another. Apart from their build, the water-
repellent coat and feet that are partially
"webbed" are also adaptations to swimming.

The Polar bear was first described as a
distinct species in 1774. It shares with the
Brown bear a common ancestor (*Ursus
etruscus*) from which both stemmed (see also
p79). That they are still closely related is
evident from the successful raising of fertile
hybrids in captivity. Recent mark-and-
recapture studies show that Polar bears

have a seasonal preference for specific geographic areas, with only limited exchange of
bears between adjacent areas. The size of
Polar bear skulls increases from east Greenland west to the Chukchi Sea, from
14.5–16in (37–41cm). It is likely therefore
that this genetic variation results from the
existence of several more or less distinct
subpopulations.

Among the ice floes of the Arctic the Polar
bear is at the top of the food chain. Polar
bears feed primarily on Ringed seals,
Bearded seals being the secondary prey
species. They also eat Harp seals and Hooded
seals and scavenge on carcasses of walrus,
Beluga whales, narwhals and Bowhead
whales. On occasion Polar bears may kill
walrus or attack Beluga whales. They occasionally eat small mammals, birds, eggs and
vegetation when other food is not available.
Polar bears catch seals in various ways. In
late April and May they break into Ringed
seal pupping dens excavated in the snow
overlying the sea ice. During the rest of the
year, seals are taken mostly by waiting at a
breathing hole or at the edge of open water.
Bears also sometimes stalk seals that are
hauled out on the ice in late spring and
summer.

Most female Polar bears first breed at five
years of age, a few at four; most breeding
males are probably older. The maximum
breeding age is not known, but reproductively active females 21 years old have been
reported. Polar bears mate in April, May and
June. One male may mate with several
females in a season, or with one. The males
locate females in heat by following their
scent. Implantation of the fertilized egg in
the uterus and its subsequent development
are delayed, resulting in a relatively long
gestation period of 195–265 days. The
pregnant females seek out denning areas in
November and December, and excavate
maternity dens, generally in drifted snow
along coastlines. The cubs are born in
December and January, the number varying
from 1 to 3, with estimates of average litter
size varying between 1.6 and 1.9 cubs. At
birth cubs weigh 21–25oz (600–700g). In
most areas, the females and cubs leave the
dens in late March and April, by which time
the cubs weigh 17–26lb (8–12kg). The
young usually remain with the mother for
28 months after birth, and the female can
breed again at about the time the young
leave her. Thus the minimum breeding
interval is usually three years.

Polar bears may travel 43mi (69km) or
more in a day; one bear that was monitored
off the coast of Alaska traveled 694mi

Conservation of a Species

The cooperation between the six countries concerned with the management and conservation of Polar bears is widely viewed as a model for conserving other species and even other natural resources.

Public concern for the Polar bear increased during the 1960s at a time of increasing human activity in the Arctic, particularly in petrochemical exploration and development. Hunting pressures also increased at this time.

The 1973 Agreement on Conservation of Polar Bears created a *de facto* high seas sanctuary by banning the hunting of bears from aircraft and large motorized boats, and in areas where they had not been previously taken by traditional means. The agreement states that nations shall protect the ecosystems of which Polar bears are a part, emphasizes the need for protection of denning and feeding areas and migration routes, and states that countries shall conduct national research, coordinate management and research for populations that occur in more than one area of national jurisdiction, and exchange research results and data. Appended resolutions request establishment of an international hide-marking system, protection of cubs, females with cubs, and bears in dens. The Convention on

International Trade in Endangered Species requires its 50-plus signatory countries to maintain records of Polar bears, or parts of bears that are exported.

The existence of more or less separate subpopulations has facilitated the management of Polar bears at a national level. Limits on hunting vary according to country. Canada allows about 600 bears to be taken annually, mainly by coastal Eskimos for meat, personal use and the sale of skins; included are a few bears (less than 15 a year) taken by licenced sport hunters guided by Eskimos. The US Marine Mammal Protection Act of 1972 transferred management authority for Polar bears from the State to the Federal government and restricted hunting to Alaskan Eskimos, who take about 100 bears each year. In Greenland (where the government shares responsibility with Denmark) Eskimos or long-time residents take 125-150 bears each year for subsistence and sale of skins. Norway has stopped nearly all hunting in the Svalbard island group because current population estimates are lower than previous ones; this bear population is shared with the USSR, where hunting has not been allowed since 1956, and the only bears taken are a few cubs (under 10 a year) for zoos.

(1,119km) during a year. Being closely associated with sea ice, most Polar bears move south in the winter as the ice extends, and north in summer as it recedes. They are most numerous in places where wind and currents keep the ice in motion, resulting in a mix of heavy ice, newly frozen ice and open water. In these conditions seals are more available to the bears. Such areas are mostly within 186mi (300km) of the coast. In some areas, bears also spend time on land, including females who use traditional maternity dens on land, and other bears who spend summer on land where ice leaves the coast or large bays.

Apart from breeding pairs and females with young, Polar bears are usually solitary. However, they occasionally congregate, and show tolerance for one another, for example at exceptionally good food sources such as a whale or a walrus carcass, where 30-40 bears have been observed, or where bears are forced ashore by ice-free conditions. Adult males are aggressive towards one another during the breeding season and also occasionally kill cubs. Some Arctic foxes that spend the winter on sea ice feed almost exclusively on the remains of seals killed by Polar bears.   JWL

AMERICAN BLACK BEAR

Ursus americanus
American black or North American black bear,
or Kermodes or Glacier bear.
One of 3 *Ursus* species.
Family: Ursidae.
Distribution: N Mexico and N California to
Alaska and across to Great Lakes,
Newfoundland and Appalachians; isolated
populations include Florida–N Gulf coast.

Habitat: forest and woodland.

Size: very variable depending
on locality and nutrition; east
of 100°W, where energy-rich
acorn and beech nut mast is
available, adult females
average 200lb (90kg) in a
155–265lb (70–120kg)
range; to the west, females
weigh 100–200lb (45–90kg),
with an average of 145lb
(65kg). Males 10–50 percent
heavier; largest recorded
males from E USA 582 and
600lb (264 and 272kg); head
to tail-tip length 4.3–5.9ft
(1.3–1.8m); shoulder height
31–37in (80–95cm).

Coat: color uniform, with wide geographical
variation (see below); sometimes a white chest
patch; black phases throughout range.

Gestation: 210–215 days.

Longevity: maximum recorded 32 years in wild
(not known in captivity).

Up to 18 subspecies recognized. Greatest
variation in coat color in W, particularly along
Pacific coast. **Kermodes bear** (*Ursus americanus
kermodei*), central coast of British Columbia;
coat can be pure white. **Blue** or **Glacier bear**
(*U. a. emmonsii*), N British Columbia coast to
Yukon; coat bluish. *U. a. altifrontalis*, SW
British Columbia; coat black and forehead high.
Cinnamon bear (*U. a. cinnamomum*), SW
Canada and W USA; coat reddish brown to
blond. *U. a. carlottae*, Queen Charlotte Islands;
large form with massive skull and black coat.
U. a. vancouveri, Vancouver Island; massive
skull, black coat. Eastern populations typically
black, eg **Eastern black bear** (*U. a. americanus*),
E and Central N America; **Newfoundland black
bear** (*U. a. hamiltoni*), with enlarged skull.
Subspecies status (especially those on
mainland) debatable; best considered races.
Differences in island races due to geographic
isolation and in mainland races due to
nutrition.

WHITE, blue and brown (and black) they
are all American black bears. The
geographical variation in coat color is very
great but because most of the range is
continuous, few clear distinctions can be
made (see BELOW LEFT).

The American black bear is much like a
small Grizzly bear. It once inhabited most
forested areas of North America, including
northern Mexico. As it is adaptable, it has
maintained much of that range in areas
where forests have been spared by man. In
extensive forested areas black bears may
overlap with the Grizzly bear, though they
are less likely to venture into the open.
Otherwise their niches are similar—both
prefer forests and are largely omnivorous,
but may take prey. Although a grizzly may
occasionally kill a black bear, there seems
to be little direct competition.

The black bear has a shorter coat (usually
without whitish hair tips) than the grizzly, a
convex outline to the head and snout, a less
sloping backline, and shorter claws that
seldom exceed 2.4in (6cm). Both young and
adults climb well.

The black bear feeds on almost any succulent, nutritious vegetation (tubers, bulbs,
berries, nuts and young shoots) and also on
grubs, carrion, fish, young hoofed mammals
or domestic stock. The search for energy-
rich food often creates conflict with bee- and
orchard-keepers. The food requirement is
some 11–18lb a day (5–8kg/day).

EATING HABITS

◀ **Mainly herbivorous,** as are most bears, the American black bear will eat meat when the chance arises, and anything else it can find. Here a bear takes the beaver it has just caught ashore to eat.

▲ **Feeding from a carcass,** in this case a steer killed in a storm.

▼ **Scavenging at a waste dump.** Scarcity of food often drives North American bears to sites where they may encounter man.

denned with the mother until April or May, then accompany her for $1\frac{1}{2}$ (sometimes for $2\frac{1}{2}$) years. Weaning occurs usually from July through September of their first year. Few American black bears approach the greatest ages recorded in the wild—23, 27 and 32 years. Sport hunting is the major cause of death, as it is with Grizzly bears. The influence of food availability on breeding is as significant as in grizzlies (see Life in the Slow Lane, BELOW).

Lone females and mother-plus-young groups often establish mutually exclusive home ranges of 1–36sq mi (2.5–94sq km); male home ranges overlap and are 5–6 times larger. Black bears are promiscuous. The female vigorously defends her litter, which may have more than one sire. As in grizzlies, a female may abandon a singleton cub—a female that carries on caring for a single cub for two years in the end rears fewer cubs than if she abandons the cub, breeds the next year and produces three young.

The American black bear suffers persecution and hunting pressures like the grizzly, but its intelligence, more secretive nature and better reproductive potential have allowed it to survive over a wider range and in greater numbers than the North American grizzly. No subspecies is endangered, although the Glacier bear and Kermodes bear are rare.   FB

Northern populations den for 5–$7\frac{1}{2}$ months each year, after which they roam large areas, foraging selectively on the richest food to regain the weight they have lost in the winter. Southern males may not den.

Black bears breed in May–July. One to three near-naked cubs are born in January or February; they weigh only 7.8–10.4oz (220–295g). The cubs remain

---

Life in the Slow Lane

Apart from an occasional dash to catch prey, the pace of life for the big North American bears is slow; their growth rate is slow, lifespan long (up to 30 years) and reproductive potential low (as few as 6–8 cubs in a lifetime).

For many bears, particularly northern populations, food is scarce and slow-growing; female American black bears have to range over areas of up to 36sq mi (94sq km) to meet their needs. The winter is long and the period of activity short. Without sufficient food a female will not reproduce: if a berry crop fails, few females will produce young that year— failed implantation of the egg at the start of denning is an "efficient" means of abortion if the female is not fat enough.

Females from eastern populations of black bear become reproductive at 3–5 years, have 2–4 cubs in a litter and may reproduce every two years. Western females, with poorer forage, do not become reproductive until 4–8 years, have litters averaging 1.7 cubs and usually wait 3–4 years between litters. Potential rates of increase of black bear numbers are thus only 12–24 percent per year. For the Grizzly bear the situation is even worse: females in northern inland areas,

where forage is very poor, range over areas of up to 77sq mi (200sq km), they do not mature until 8–10 years of age, and average litter size is 1.7, with 4–5 years between litters. Better-fed coastal females mature at 4–6 years, have litters averaging 2.2 cubs and conceive every 3–4 years. Reproductive potential is thus as low as 6–16 percent per year for grizzlies.

For males, the consequences of food scarcity are equally significant. To find enough food and locate the few females available for mating, adult male black bears search vast areas up to 230sq mi (600sq km) (grizzlies range over 154–425sq mi (400–1,100sq km).

Adult males may kill young males still accompanying their mothers if the female is receptive (in estrus). Aggression results as both male and female try to maximize reproductive success (see p91).

With food scarce, yet critical for reproduction, black bears and grizzlies are attracted to dumps, beehives, livestock, or the bait stations of hunters, which makes them more vulnerable to man. With such low rates of reproduction, these populations can sustain only a very low rate (0–5 percent per year) of additional "unnatural" deaths. Each bear killed by man is a significant loss.

SMALL BEARS

Four species in 4 genera
Family: Ursidae.

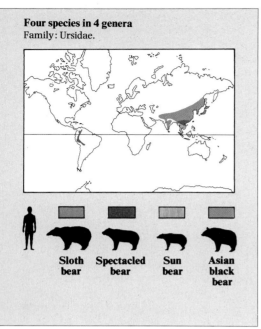

Sloth bear   Spectacled bear   Sun bear   Asian black bear

THE four smaller bears are more southerly in distribution than the *Ursus* species. Each is placed in a separate genus.

The **Sloth bear** differs markedly from other bears. With its long curved claws it can hang sloth-like upside down. Its naked and flexible lips and long snout, nostrils that can be closed, hollowed palate, and lack of two inner upper incisor teeth are specialized feeding adaptations unique in the bear family. The coarse coat is usually black, but often mixed with brown, gray or rusty-red.

Sloth bears are primarily nocturnal. They are omnivorous, eating insects, grubs, sugarcane, honey, eggs, carrion, fruits and flowers. When feeding on termites, a Sloth bear breaks open a termite mound with its claws, and uses its lips and long tongue as a tube, first to blow away dust, then to suck up the prey. The claws are equally useful when foraging in trees for fruits and flowers.

Northern populations breed in June,

southern populations all year round. Seven months later 1–3 (usually 2) young are born in a ground shelter. The Sloth bear does not become dormant, but dens for seclusion and protection. Cubs leave the den after 2–3 months but accompany the mother for two, possibly three, years. Sloth bears are reported to have only one mate.

The **Spectacled bear** of South America is descended from ancestors which entered the continent from North America some 2 million years ago. The markings around each eye vary considerably between individuals. The thick coat is otherwise of uniform color.

The Spectacled bear lives in a variety of habitats and altitudes from 650–14,000ft (200–4,200m). Although it prefers humid forests, it makes use of grasslands above 10,500ft (3,200m) and lower lying scrub deserts—all habitats threatened by human encroachment.

It is a good climber, commonly foraging in trees in search of succulent bromeliad "hearts," petioles of palm fronds, and fruits such as figs in the forests and cactus in the desert. Fruit-bearing branches, broken while foraging, may be pulled together as a platform or nest which is sometimes used as a day-bed. Although primarily herbivorous, the Spectacled bear also feeds on insects, carrion, occasionally domestic stock and, reportedly, young deer, guanacos and vicuñas. The Spectacled bear appears to be active throughout the year. Litters comprise 1–3 (usually 2) small cubs of 10.5–11.5oz (300–325g).

The **Sun bear** or Malayan sun bear is the smallest member of the bear family. It is also the one with the shortest and sleekest coat—perhaps an adaptation to a lowland equatorial climate.

Although it inhabits both lowlands and highlands, the Sun bear is primarily a forest dweller, resting and feeding in trees in tropical to subtropical regions of Southeast Asia (Borneo, Sumatra, Malay Peninsula, Kampuchea, Vietnam, Laos, Burma, and possibly southern China).

Relatively low weight, strongly curved claws, and large paws with naked soles help to make the Sun bear an adept climber. It is primarily nocturnal, frequently resting or sunbathing during the day on a platform of broken branches several feet above ground level. It is omnivorous, eating tree fruits, succulent growing tips of palm trees, termites, small mammals and birds, and can cause significant damage in cocoa and coconut plantations.

Sun bears may mate at any time of year. They do not become dormant. The young,

▲ **The four smallest bears** all occur in the tropics. (1) The shaggy-coated Sloth bear makes good use of its long curved claws and flexible snout to forage, either on the ground for termites and grubs or in trees. The Sun bear in the foreground (2) is the smallest bear; here it licks termites from the mound it has broken open. The Spectacled bear (3), shown climbing a tree in search of fruit, is the only South American bear. The Asian black bear (4) is mainly herbivorous but may, as here, take carrion.

◀ **Performing bears,** once common, are a rare sight today. This Indian Sloth bear must be one of the last "trained" members of what is a threatened species.

usually two, each weighing 10.5–12oz (300–340g), are born in seclusion on the ground. Sun bears are thought to have only one mate. Their cautious nature and small size make them, for man, the least dangerous of bears, and for this reason locals sometimes keep them as pets.

The **Asian black bear** inhabits forest and brush cover (in places, together with the Brown bear) from Iran through the Himalayas to Japan.

The Asian black bear is somewhat smaller than the American black bear which it resembles in habits and with which it may share a common ancestor (see p79). In addition to the typical jet black coloration, brown and reddish brown individuals also occur. The generic name *Selenarctos*, meaning "moon bear," derives from the white

chest mark. There is often a "mane" of longer hairs at the neck and shoulders. The very prominent ears are rounded.

The species is omnivorous, feeding mainly on plant material, especially nuts and fruit, but also ants and larvae. It is a good climber and frequently forages in trees and on succulent vegetation on avalanche slopes. In summer it sleeps or rests on tree platforms built of branches broken while feeding.

Asian black bears may seek out cultivated crops or domestic livestock, although they tend to avoid human contact. In Japan they cause serious damage to forest plantations by feeding on the living tissue of tree bark. Only northern populations den consistently in winter. Females with their young (usually one or two) leave the den in May and stay together about two years.   FB

Abbreviations: HTL = head-to-tail-tip length; HT = height; TL = tail length; wt = weight.
[I] Threatened, status indeterminate.
[V] Vulnerable.   [*] CITES listed.

Sloth bear [I]
Melursus ursinus

E India and Sri Lanka; lowland forests. HTL 5–6ft; shoulder HT 24–35in; wt 198–253lb, occasionally up to 298lb; males larger than females. Coat: long and shaggy, usually black, with white to chestnut U- to Y-shaped chest mark. Gestation: about 210 days. Longevity: to 30 years in captivity (not known in wild).

Spectacled bear [V]
Tremarctos ornatus

Andes, from W Venezuela to Bolivia; various habitats, but prefers humid forests. HTL (males) 4–7ft; shoulder HT 27–35in; wt commonly 286lb but up to 440lb; females much smaller, 77–143lb. Coat: black or brown-black, with white to tawny "spectacles" sometimes extending to chest. Gestation: 240-255 days. Longevity: to 20–25 years in captivity (not known in wild).

Sun bear [*]
Helarctos malayanus
Sun bear or Malayan sun bear.

SE Asia; primarily forests. HTL (males) 3–4.5ft; shoulder HT 27in; wt 59–143lb, females about 20 percent less. Coat: deep brown to black, often a whitish or orange chest mark; light fur (usually grayish or orange) on the short, mobile muzzle. Gestation: about 96 days. Longevity: not known.

Asian black bear [*]
Selenarctos thibetanus
Asian black bear or Himalayan black bear.

Iran to Japan; forests and brush cover. HTL 4–5.5ft; wt: males 110–264lb, females 92–154lb. Coat: long, jet black with purplish sheen; white crescent on chest, some white on chin. Gestation: not known. Longevity: to 24 years in wild (not known in captivity).

THE RACCOON FAMILY

Family: Procyonidae
Seventeen species in 8 genera.
Distribution: N, C and S America; pandas in Asia.

Habitat: very diverse, from cool temperate to tropical rain forest; Common raccoon in urban and agricultural areas.

Size: head-body length from 12–15in (31–38cm) in the ringtail to 59in (150cm) in the Giant panda; weight from 1.8–2.4lb (0.8–1.1kg) in the ringtail to 220–330lb (100–150kg) in the Giant panda.

Raccoons and coatis
(subfamily Procyoninae)
Fifteen species in 6 genera.
Raccoons Six species of *Procyon*.
Coatis Three species of *Nasua*, 1 *Nasuella*.
Olingos Two species of *Bassaricyon*.
Ringtail and **cacomistle** Two species of *Bassariscus*.
Kinkajou *Potos flavus*

Pandas (subfamily Ailurinae)
Two species in 2 genera.
Giant panda *Ailuropoda melanoleuca*
Red panda *Ailurus fulgens*.

THE raccoon family contains just 17 species, but its members show a remarkable diversity in form and ecology. This diversity is reflected in the scientific debate about its classification, which still rages today. Here we take the view that there are two subfamilies, the Ailurinae (the herbivorous pandas) and the Procyoninae (the other 15 omnivorous species). Most controversy surrounds the position of the two panda species. Some consider them so distinct as to belong to a separate family, the Ailuropodidae, while others retain the Red panda alone in the Ailurinae and place the Giant panda in the Ailuropodidae or even with the bear family, Ursidae.

The raccoon family is descended from the dog family, Canidae. Recognizable fossil *Bassariscus* have been found from 20 million years ago, a time when Europe and North America were one continent. When the continents separated, the family split, with procyonines remaining in the New World and the ailurines, resembling the Red panda, in the Old World.

Procyonids, except the Giant panda, are small, long-bodied animals with long tails. The kinkajou is uniformly colored, but the others have distinctive markings, ringed tails (except the Giant panda) and facial markings that vary from the black mask of the raccoons to white spots in the coatis and

▶ **Feeding techniques** and other features of members of the family Procyonidae. (1) Coati grubbing for insects. (2) Ringtail eating a lizard. (3) Head of a cacomistle. (4) Kinkajou licking nectar from a flower while holding on with its prehensile tail. (5) Tail of olingo, which is bushy not prehensile. (6) Giant panda eating bamboo shoots.

▼ **The unmistakable fox-like face** and markings of a Common raccoon at the entrance of its den. Like most procyonids, raccoons are generally active at night.

Skulls of Procyonids

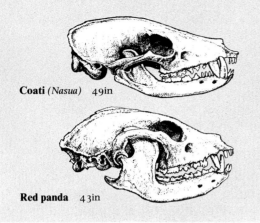

Coati (*Nasua*) 49in

Red panda 43in

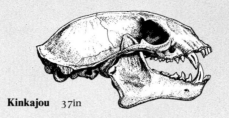

Kinkajou 37in

The teeth of procyonids are generalized, as befits omnivores. The typical dental formula (see p12) is I3/3, C1/1, P4/4, M2/2 = 40 but this varies with species. The kinkajou has only three premolars above and below. The Red panda has three premolars above and four below; and the Giant panda three or four above and three below plus an extra molar below. Only the cacomistle has well-developed carnassials. In raccoons the carnassials are unspecialized and the molars flat-crowned. In coatis the molars and premolars are high-cusped as adaptations to a more insectivorous diet: the canines are long and blade-like and may be used for cutting roots while digging.

cacomistle. The Giant panda is white, with black legs, shoulders, ears and eyepatches. The Red panda is red with a white face.

The feet of procyonids have five toes and the animals walk partly or wholly on the soles of their feet (plantigrade gait). The Giant panda has an extra digit that functions like an opposable thumb, as does the Red panda, although the digit is much smaller. Claws are nonretractile, except that ringtails and Red pandas have semi-retractile claws on their forepaws. Kinkajous have a prehensile tail and long tongue used in feeding. The muzzle is usually pointed, although the kinkajou has a short muzzle and coatis a long, flexible snout. The cacomistle has unusually large ears to help it locate prey.

The small procyonids can live 10–15 years in captivity, but rarely more than seven in the wild. Females usually breed in the spring of their first year, males from their second year on. Gestation varies from 63 days in raccoons to five months in Giant pandas. The young weigh about 5.3oz (150g) and are poorly developed at birth, even in the Giant panda. In most species there are 3–4 young in a litter, but Red pandas have only one or two and Giant pandas and kinkajous usually only one. Females bear their litters in dens or nests and provide all of the parental care.

All the procyonids are nocturnal, except coatis, which are active by day, and the Giant panda, active at twilight. Except possibly ringtails, they are solitary but not territorial. Neighbors sometimes fight, but usually simply avoid each other. Kinkajous are tolerant of each other and of olingos in fruit trees. Females are accompanied by their young for a few months to a year. Coati females, unlike the others, live in social groups.

Most procyonid species are thriving, with exception of the Barbados raccoon, which may be extinct, and the Giant panda.   JKR

RACCOONS

All six species of the genus _Procyon_
Family: Procyonidae.
Distribution: N, C and S America.

Size: head-body length 22in
(55cm); tail length 10in (25cm).
Weight usually 11–18lb (5–8kg),
sometimes up to 33lb (15kg),
females being about 25 percent
smaller than males.

Gestation: 63 days.

Longevity: not known in wild (over 12 years
recorded in captivity).

Common raccoon
Procyon lotor
S Canada, USA, C America; introduced in parts
of Europe, Asia. Commonest species, occupying
diverse habitats. Coat: usually grizzled gray but
sometimes lighter, more rufous (albinos also
occur); tail with alternate brown and black
rings (usually 5); black face mask accentuated
by gray bars above and below, black eyes and
short, rounded, light-tipped ear pinnae.

Tres Marías raccoon
Procyon insularis
María Madre Island, Mexico. Coat shorter,
coarser, lighter-colored than _P. lotor_.

Barbados raccoon
Procyon gloveranni
Barbados. Coat darker than _P. lotor_. Very rare.

Crab-eating raccoon
Procyon cancrivorus
Costa Rica south to N Argentina. Coat shorter,
coarser, more yellowish-red and with less
underfur than _P. lotor_; hair on nape of neck
directed forward. Tail longer than _P. lotor_.

Cozumel Island raccoon
Procyon pygmaeus
Cozumel Island, Yucatán, Mexico. Coat lighter
than _P. lotor_. The smallest raccoon, often only
6.6–8.8lb (3–4kg).

Guadeloupe raccoon
Procyon minor
Guadeloupe. Coat paler than _P. lotor_.

RACCOONS are mischievous animals, notorious as crop marauders, garbage bandits and escape artists. Physically, they are quite unmistakable: a fox-like face with a black mask across the eyes, a stout cat-like build and a ringed tail. Young "coons" make enchanting pets, but when adult their insatiable curiosity, destructive nature and general untrustworthiness can try the most devoted of owners.

The popular name "raccoon" originated from a North American Indian word _aroughcan_ or _arakun_ (roughly translated as "he who scratches with his hands"). The species epithet _lotor_ refers to this species' habit, in captivity, of apparently "washing" food and other items. The term "washing" is in fact a misnomer. In the wild, similar actions of rubbing, feeling and dunking, using their highly dextrous and sensitive front paws, are associated with location and capture of aquatic prey, such as crayfish and frogs. Whether these actions are simply investigative or intended to rid the prey of distasteful skin secretions is not known. However, the behavior is innate, and captive animals unable to give vent to their tendencies naturally will relieve their frustration by simulating the actions on any prey-like object, even in the absence of water.

In most areas raccoons forage at night near streams or marshy areas, where frogs, crayfish, fish, birds and eggs are sought. However, upland areas are also frequented in search of fruit, nuts and small rodents. They also consume insects and are even known to eat earthworms. Fresh corn appears to be a particular delicacy and since raccoons harvest corn before farmers, they are considered a nuisance in many areas. Raccoons have no aversion to living near humans and sometimes seek shelter in barns, sheds and other buildings. They occur in many urban areas, especially near parks or ravines, and night raids on garbage bins often annoy people.

The Common raccoon has been extending its range northwards in recent years

▲ **Begging for food.** Common raccoons normally forage on their own for prey, but where food is potentially abundant they will congregate. This group is begging beside a road, where they know passers-by will give them scraps.

◀ **Peacefully suckling her young,** a Common raccoon uses her dextrous paw to control one of her cubs. Raccoons den in ground burrows and in or near human habitation as well as in trees.

▼ **The Crab-eating raccoon** from South America is a good climber, like other species, with a longer tail and shorter, coarser and yellower hair than the Common raccoon. Crabs form only a minor part of its diet.

Similar ground dens were also used by Red foxes, Striped skunks, porcupines, and their likely originators, woodchucks. Other den sites are brush and log piles, barns, and even the attics of old houses.

In Common raccoons, the mating season is late January and early February. Raccoons are probably polygynous, with one male searching the ranges of, and visiting, two or three females. Home ranges of adult females overlap, but adult males seldom occur in the same range except near seasonally abundant food supplies. More than one male may be found in different parts of a female's range; scars seen on adult males suggest that aggressive encounters may occur in the competition to find a mate. Year-round ranges of raccoons vary from 125–12,350 acres (50–5,000 hectares) depending on population density and the abundance of prey, but most raccoons regularly forage over about 800 hectares (2,000 acres). Movements by adult females are quite restricted in April, when 3–7 young are born. There are records of raccoons born during the late summer months, which suggests that some raccoons can mate much later than midwinter. This may occur when severe weather, preventing traveling, coincides with a female's first menstruation and mating occurs during a later period in heat. Juveniles generally forage with or near their mother for the first year. In southern areas, dispersal (if any) of juveniles occurs in the autumn, but in northern areas not until the spring when the next litter arrives. Males disperse more frequently and over longer distances than females. Northern raccoon populations also have individuals with longer and thicker coats, heavier weights, fewer breeding yearlings, larger litters, mutually exclusive territories among males, and winter denning.

The sport of "coon hunting" is prevalent during September to December each year, especially in eastern North America. High fur prices ($25–$50 each) during the 1970s intensified the interest in trapping and hunting raccoons, although night-time hunting with specially bred hounds has a long tradition, second only, perhaps, to that of English fox hunting. Over four million raccoons are harvested annually either this way or by trapping, and many others die each year on roads. Parasites, diseases (such as distemper) and malnutrition after severe winters also cause mortality. The Common raccoon is the major rabies vector in the southeastern United States, and in recent years raccoon rabies has spread northward to Virginia and Maryland.   DRV

coincident with increasing land clearance for agriculture and a gradually warming climate. It has been successfully introduced to Europe and Asia, where there are wild populations in some areas. The other species are restricted in distribution. Most unusual is the semiaquatic Crab-eating raccoon, which was once placed in a separate genus, *Euprocyon.* Crabs are not its principal food, but it does prey on several types of animal found in or near its aquatic habitat.

In the northern United States and southern Canada, the Common raccoon becomes very inactive during the winter months, although it is not a true hibernator. They will remain in the same den for a month or more unless temperatures rise above freezing at night. Communal denning is common and a female and her offspring of the year often hunt independently but den together. Up to 23 raccoons have been reported in a single den, but more than one adult male in a den is rare. As far as is known most raccoons that den together are relatives. Since the Common raccoon is an excellent climber and is often seen in trees when pursued, many people consider that large hollows in trees are the preferred dens for raccoons. Although females may prefer arboreal protection for their young in the spring, recent radio-tracking studies in Ontario and the northern prairies have shown that raccoons often use ground burrows. Even when there were many tree dens available, raccoons had more dens in the ground and spent more time in such sites.

COATIS

Four species in 2 genera
Family: Procyonidae.
Distribution: Southern N America, C and
S America.

Habitat: wide-ranging, including tropical
lowlands, dry high-altitude forests, oak forests,
mesquite grassland and on the edge of forests.

Gestation: 77 days.

Longevity: 7 years (to 14 years in captivity).

Ringtailed coati
Nasua nasua
Forests of S America, E of Andes, S to
N Argentina and Uruguay.

Size: head-to-tail-tip length
32–51in (80–130cm), somewhat
more than half being tail; weight
8.8–12.3lb (4.0–5.6kg) (male):
7.7–9.9lb (3.5–4.5kg) (female).
Coat: tawny-red with black face; a
small white spot above and below
each eye and a large one on each
cheek; white throat, belly; black
feet, black rings on tail.

White-nosed coati
Nasua narica
SE Arizona, Mexico, C America, W Colombia
and Ecuador.
Size: as Ringtailed coati. Coat: gray or brown
with silver grizzling on sides of arms and a
white band round the end of the muzzle; other
facial markings, throat, belly, feet and tail as in
Ringtailed coati.

Island coati
Nasua nelsoni
Cozumel Island, near Yucatán Peninsula,
Mexico.
Size: head-to-tail-tip length 28–32in
(70–80cm), tail shorter than body. Coat:
shorter, softer and silkier than the White-nosed
coati, otherwise similar.

Mountain coati
Nasuella olivacea
Mountain forests of Ecuador, Colombia and
W Venezuela.
Size: head-to-tail-tip length 28–32in
(70–80cm), tail shorter than the body. Coat:
olive brown, with black muzzle, eye rings and
feet; tail with black rings.

A long, ringed tail, often held erect above the body, and a highly mobile, upturned and elongated snout are characteristic physical features of coatis. Socially, they are also very distinctive—the males are solitary, while females live in highly organized groups, with individuals often caring for each other's young. Unlike their relatives, they are mainly active in the daytime.

Coatis have strong forelimbs and long claws. They can reverse their ankles and descend trees head first. Their long tail is used as a balancing rod while climbing.

Coatis are primarily insectivorous, but are also fond of fruit. They forage with their nose close to the ground, sniffing in leaf litter and rotting logs on the forest floor for beetles, grubs, ants and termites, spiders, scorpions, centipedes and land crabs, which are excavated with their forepaws. They also occasionally catch frogs, lizards and mice, and unearth turtle and lizard eggs. They eat

When Is a Coati a Coatimundi?

Female coatis and their young associate in bands of 5–12 individuals, but adult males are solitary. This difference at first confused biologists, who described the solitary males as a separate species. The use of the name "coatimundi"—meaning "lone coati" in Guarani—for this "species" reflects the same error.

The social bonds between adult females, often shown by mutual grooming (1), bring benefits to all the members of a coati band, but mostly to the juveniles. The adults and subadults surround the juveniles, and also watch for predators (2). Adults will cooperate to chase a predator away vigorously. Juveniles also get nearly as much grooming (3) from their mother's allies as they do from her and are even allowed to nurse. They also spend much time playing together (4).

One theory to explain this behavior was that allied females must be genetically closely related in order to have any interest in caring for each other's young. However, recent fieldwork shows that normal cooperative relationships occur frequently among unrelated females. The term "reciprocal altruism," essentially a formal description of friendship, is used to describe these relationships. They take time to develop, but once established they allow "friends" reciprocally to protect and tend each other's offspring. As a result, females can safely bring their young down from the nest 6–10 weeks earlier than can other procyonid mothers and can devote more of their attention to searching for food than they could if they had to protect their young themselves. During the nesting season, lone females devote about 18 percent of their foraging time to vigilance, whereas after rejoining the band the proportion is only about 10 percent.

ripe fruits both on the ground and high in trees. Males forage quietly alone, and catch more lizards and rodents than females and young, who forage in groups.

In the larger species, females mature in their second year, males in their third. Females chase males away during most of the year, because they sometimes kill juveniles, but in February and March they become more tolerant. Usually, a single male, normally the most dominant one at the center of the band's range, breeds with each band of females. He gradually wins his way into the band by grooming its members and behaving submissively. Mating occurs in a tree. Soon afterwards the male is expelled from the band by the females, who again become aggressive. About 3–4 weeks before giving birth, females separate from their bands to build nests in trees. Platform nests may be assembled from sticks, but often a large palm tree is chosen, the crown of which makes a natural nest with little modification. Coatis bear litters of 3–5 poorly-developed young weighing only 3.5–6.4oz (100–180g) and keep them on the nest for 5–6 weeks, leaving only for brief periods to forage. By late May, when the young weigh about 18oz (500g), mothers bring them down from the nest and they rejoin bands. Soon afterwards the male that bred with the band often joins it for a few minutes on several successive days and grooms with all its members, including the new juveniles. These encounters apparently allow males to recognize their own offspring in order to avoid preying on them later.

The home ranges of bands are about 0.6mi (1km) in diameter and each band's range is completely overlapped by the peripheries of its neighbors' ranges, although in any area one band is encountered much more often than any other. Bands usually tolerate each other when they meet, sometimes even foraging or grooming together. Each band's range includes the ranges of several adult males. New bands usually come about through the splitting of a band and the friendly relationships between neighbors reflect a continuation of earlier social bonds. Nonetheless, each band's membership is stable and distinct. Males mark their ranges by dragging their abdomens on branches and often fight when they meet, but their ranges do overlap.

Coatis interact little with humans, except for occasionally raiding crops planted near forest.

JKR

▲ **Tail-up posture** of the White-nosed coati which is often displayed when excavating food items.

◄ **Long, mobile snout and strong front claws** are tools of the trade for the Ring-tailed coati, which snuffles around in the litter on the forest floor for insect prey and excavates the surface soil to grub out lizards and spiders from their burrows.

GIANT PANDA

Ailuropoda melanoleuca [R]
Giant panda, Panda, Panda bear or Bamboo bear.
Sole member of genus.
Family: Procyonidae.
Distribution: Szechuan, Shensi and Kansu provinces of C and W China.

Habitat: cool, damp bamboo forests at altitudes of 8,500–11,500ft (2,600–3,500m).

Size: shoulder height 27–32in (70–80cm); can measure about 67in (170cm) when standing; weight 220–330lb (100–150kg), males being about 10 percent larger than females.

Coat: ears, eye patches, muzzle, hind limbs, forelimbs and shoulders black, the rest all white.

Gestation: 125–150 days.

Longevity: unknown in the wild (over 20 years in captivity).

[R] Rare.

▼ **Boldly marked,** "cuddly" and "human," the Giant panda is at the same time one of the rarest and one of the best known of animals.

THE Giant panda's rise to fame has been rapid. It was discovered only in 1869 and the first captive animal was not brought to the West until 1937. In recent years they have become precious diplomatic gifts, presented to only a few fortunate and highly favored countries. The attraction derives not only from their scarcity, but also from their unique combination of endearing traits, such as apparent cuddliness, short face, bold color pattern and human-like ways of sitting and eating. The Giant panda's scarcity and popularity led to its adoption as the symbol of the World Wildlife Fund.

The Giant panda is often referred to simply as "the panda" and sometimes misleadingly as the Panda bear or Bamboo bear. Its taxonomic position is debatable (see p90). Giant pandas are difficult to observe in their natural habitat, so little is known about them in the wild.

Giant pandas are bear-like in shape with striking black and white markings. They have what is in effect an extra digit on their forepaws. One of the wrist bones, the radial sesamoid, is much enlarged and elongated; it is used like a thumb to oppose the rest of the digits. This enables the Giant panda to grip slender pieces of food in its forepaws—notably the bamboo that is by far the largest item of the diet. Much of the bamboo stem ingested is passed out relatively unchanged in the many large feces about 10 hours later and digestion of food is clearly inefficient. The animal usually sits upright while feeding, holding the food in a forepaw. Small food items may be eaten directly off the ground. Pandas in the wild also eat bulbs, grasses and occasional insects and rodents.

Full sexual maturity is reached at 4–5 years old for females and probably a couple of years later for males. The female is apparently receptive to the male for only one period in the year, in April–May. It is not known how male and female find one another at this time, but voice and scent are probably important—both calling and scent marking increase around the time of mating. Mating itself is brief and unelaborate. The litter size is one, two or occasionally three, but apparently only one cub can be reared, for they are small and helpless, weighing 3–5oz (100–150g), and blind at birth. They are born in a sheltered den. The eyes open at 1½–2 months, the cub is mobile at three months, is said to be weaned at six months and may be independent at twelve.

Giant pandas are largely solitary in the wild and probably have a territorial system like that of the leopard (see pp38–39). Scent marking is common, especially in males, by

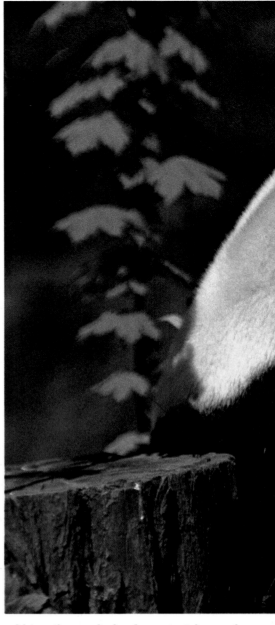

rubbing the anal glands against large objects. There are a few simple vocal signals, used at short range. A louder whinny-like bleating carries a few hundred meters.

Giant pandas are one of the rarest of mammals—there are probably fewer than 1,000 in the wild, perhaps even less than 500. They are not persecuted, and indeed they are protected by sentiment and tradition, and by law in mountain forest reserves. These areas are generally both inaccessible and inhospitable, so there should be little human pressure on them for cultivation or settlement. The Giant panda is listed as rare in the IUCN *Red Data Book*, rather than as endangered. In recent years, there has been a widespread increase in mortality following the flowering, seeding and dieback of the bamboo *Sinarundinaria nitida*, which takes place only about every 100 years. Obviously the species has encountered and overcome this phenomenon over the centuries, and certainly Giant pandas are able to supplement their bamboo

▲ **A hidden "thumb"** provided by the enlargement of one of the wrist bones allows the Giant panda to grip bamboo stems. Mostly leaves and slender stems are eaten, but Giant pandas can also cope with stems up to about 1.5in (40mm) in diameter. The carnassial teeth are adapted to both slicing and crushing and there is also an extra lower molar.

diet with other foods if necessary and if these are available. But the increase in human population and settlement around the areas of Giant panda habitat mean that the animals can no longer move to and restock other isolated patches of suitable habitat. Human intervention to protect the species may become necessary.

Small isolated populations are always at risk, so a breeding captive population is a vitally important backup. Although a few Giant pandas have been kept in captivity for many years, breeding has been poor. The first cub actually reared was born in 1963 in Peking Zoo, and since then there have been only a dozen other successes. The fundamental problem is that captive Giant pandas are so rare and highly prized. Most nations or zoos have only a pair or a single animal, and are not prepared to send them to places where they might breed better. There have been a variety of problems; they include a female too accustomed to humans to accept a male panda (London), a male who does

not adopt the right mating posture (Washington), a sick female (London), a sick male (Madrid), a pregnant female who died (Tokyo), and males who either are aggressive towards or not aroused by females.

One solution to the problem of incompatibility between the sexes is artificial insemination. The technique has produced a few results in China since 1978, but it is difficult to determine the success rate. It is of value only in cases where the female becomes fully fertile but where natural mating cannot take place. The method depends on reliable ways of immobilizing Giant pandas, and on development of techniques to store semen.

The survival rate of Giant panda cubs in captivity is low, as it probably is in the wild. A high proportion of litters consists of twins; the mother only attempts to rear one of them, and ignores the other. In principle, the reproduction rate in captivity could be considerably increased if methods could be developed for handrearing these abandoned cubs; attempts so far have failed. BCRB

OTHER PROCYONIDS

Five species in 3 genera
Family: Procyonidae.

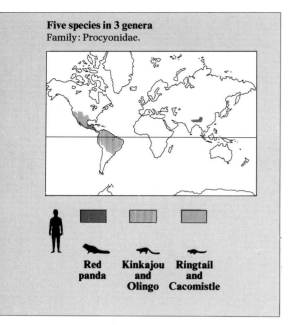

Red
panda

Kinkajou
and
Olingo

Ringtail
and
Cacomistle

▼ **The Red panda** is fairly nocturnal, as indicated by the well-developed whiskers. The soft, deep-red fur and white face markings are distinctive.

A MONG the least known of procyonids are the nocturnal species of Central and South America, and the Red panda of Asia. Although now overshadowed by the fame of the Giant panda, the **Red panda** was for 50 years the only panda known to man. It has distinctive red fur and is more widespread than the Giant panda. Like its much larger relative, the Red panda has an extra "thumb" (the enlarged radial sesamoid), although it is less well developed.

Red pandas have a varied, mainly vegetarian diet—fruit, roots, bamboo shoots, acorns and lichens are reportedly eaten. In captivity, they readily consume meat, so it is likely that in the wild they eat some insects or carrion. They are excellent climbers and probably forage mostly in trees.

The birth season is from mid-May to mid-July. Although Red pandas normally have a single period in heat, there are suggestions that either they have several or they exhibit delayed implantation of the fertilized egg. One to four young (commonly two) are born, fully furred but blind and helpless, in a hollow tree. They are weaned at around five months and become sexually mature at 18–20 months. Males take no part in rearing their young.

Adult Red pandas are fairly nocturnal, are believed to be solitary and are probably territorial. Males in particular scent mark using their anal glands. They also use regular defecation sites, which may serve as territorial markers. Both subspecies of Red panda are scarce and are declining. A small amount of illegal hunting takes place, but much more serious is the extensive deforestation which has accompanied the increase in local human populations. *A. f. styani* is reasonably well protected in reserves in China. There are about 140 Red pandas in captivity, about half of them captive-bred.

The **kinkajou** and the **olingos** are strikingly similar in external appearance and habits. All have long bodies, short legs, long tails and are nocturnal fruit-bearers; the species are sometimes even found foraging together, usually several kinkajous and one or two olingos—but they are difficult to tell apart.

A closer examination reveals important differences. Kinkajous are slightly larger than olingos, have foreshortened muzzles

Abbreviations: HBL = head-body length; TL = tail length; wt = weight.
[*] CITES listed.

Red panda [*]

Ailurus fulgens
Red or Lesser panda.

Himalayas to S China. Favors remote, high-altitude bamboo forests. HBL 20–24in; TL 12–20in; wt 7–11lb. Coat: soft, dense, rich chestnut-colored fur on the back; limbs and underside darker; variable amount of white on face and ears. Gestation: 90–145 days. Longevity: up to 14 years. Subspecies: 2; *Ailurus fulgens fulgens*, Himalayas from Nepal to Assam. *Ailurus fulgens styani*, N Burma and S China.

Kinkajou

Potos flavus

S Mexico to Brazil, in tropical forest. HBL 16–22in; TL 16–22in; wt 3–6lb. Coat: short, uniformly brown. Gestation: 112–118 days. Longevity: up to 23 years in captivity.

Olingo

Two species of *Bassaricyon*.

Bassaricyon gabbii in C America and northwestern S America; *B. alleni* in Amazonia. In tropical rain forest at about 5,900ft. HBL 16–18in; TL 17–19in; wt about 3lb. Coat: gray-brown, long, loose hair with blackish hues above, yellowish below and on insides of the limbs; yellowish band across neck to back of ears; tail with 11–13 black rings, often indistinct. Gestation: 73–74 days. Longevity: more than 15 years in captivity. *Bassaricyon lasius* (Costa Rica) and *B. pauli* (Panama) are probably subspecies of *B. gabbii*, and *B. beddardi* (British Guiana) a subspecies of *B. alleni*.

Ringtail

Bassariscus astutus
Ringtail, Civet cat, Miner's cat, or Ring-tailed cat.

W USA, from Oregon and Colorado south and throughout Mexico. Dry habitats, especially rocky cliffs. HBL 12–15in; TL 12–17in; wt 2–3lb. Coat overall gray or brown; white spots above and below each eye and on cheeks.

Cacomistle

Bassariscus sumichrasti

C America. Dry forests. HBL 15–20in; TL 15–21in; wt 2lb. Coat as above.

▲ ◄ **The kinkajou** of Central and South American forests is primarily a fruit-eater. It uses its long tongue to probe the nectaries of flowers and when obtaining honey.

◄ **Solitary and nocturnal** denizen of Central American forests, the cacomistle has tapered ears, while the ears of its very similar northern relative, the ringtail or Miner's cat, are rounded.

and short-haired prehensile tails; olingos have long muzzles and bushy non-prehensile tails. Kinkajous also have a long extrudable tongue, possibly used to reach nectar and honey, and lack one premolar. They also lack anal sacs, instead of which they have scent glands on the chest and belly. Kinkajous eat only fruit and other sugary foods, while olingos also eat large insects, small mammals and birds. In all, the olingos are normal procyonids, while kinkajous are considered by some to merit subfamily status.

Very little is known about the life-style of these elusive animals. Kinkajous breed throughout the year; they produce a single young each year. Large numbers will congregate in fruiting trees, but whether these groups remain together is doubtful.

The **ringtail** is a graceful carnivore which was often reared as a companion and mouser in prospectors' camps in the early American West—hence the name Miner's cat. Although one of the smallest procyonids, it is the most carnivorous. Both the ringtail and its slightly larger relative the **cacomistle** have dog-like teeth which also reflect their predatory nature. The cacomistle spends much more time in trees than the ringtail. Both species have long legs, lithe bodies and long, bushy, ringed tails. They have fox-like faces, and their ears are larger than in other procyonids: the ringtail's are rounded while the cacomistle's are tapered. The ringtail has semiretractile claws, the cacomistle's are nonretractile.

Both prey heavily on lizards and on small mammals up to the size of rabbits, but they also eat large insects, fruit, grain and nuts. They are strictly nocturnal and although fairly common are rarely seen. Both are solitary and have home ranges of more than 250 acres (100 hectares). Generally, only one male and one female are found in any area. JKR

THE WEASEL FAMILY

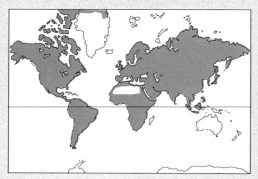

Family: Mustelidae
Sixty-seven species in 26 genera.
Distribution: all continents except Antarctica and Australia (but introduced into New Zealand).

Habitat: from Arctic tundra to tropical rain forest, on land, in trees, rivers and the open sea.

Size: smallest is the Least weasel: head-body length 6–8in (15–20cm), tail length 1–1.5in (3–4cm), weight 1–2.5oz (30–70g). Many species under 2.2lb (1kg). Giant otter head-body length 38–48in (96–123cm), tail length 18–26in (45–65cm), up to 66lb (30kg); the Sea otter may reach 100lb (45kg). Males larger than females, often considerably so.

Weasels and allies (subfamily Mustelinae)
Thirty-three species in 10 genera, including **European common weasel** and **Least weasel** (*Mustela nivalis*), **ermine** (*M. erminea*), **European polecat** (*M. putorius*), **Long-tailed weasel** (*M. frenata*), **kolinsky** (*M. sibirica*), **American mink** (*M. vison*), **Stone marten** (*Martes foina*) or Beech or House marten, **Pine marten** (*M. martes*), **sable** (*M. zibellina*), **fisher** (*M. pennanti*), **wolverine** (*Gulo gulo*).

Skunks (subfamily Mephitinae)
Thirteen species in 3 genera, including **Spotted skunks** (*Spilogale*) and **Striped skunk** (*Mephitis mephitis*).

Otters (subfamily Lutrinae)
Twelve species in 6 genera, including the **North American river otter** (*Lutra canadensis*), **Eurasian otter** (*L. lutra*), **Spot-necked otter** (*Hydrictis maculicollis*), **Oriental small-clawed** and **Cape clawless otters** (*Anonyx cinerea* and *A. capensis*), **Giant otter** (*Pteronura brasiliensis*) and **Sea otter** (*Enhydra lutris*).

Badgers (subfamily Melinae)
Eight species in 6 genera, including the **Eurasian badger** (*Meles meles*), **American badger** (*Taxidea taxus*) and **Ferret badgers** (*Melogale*).

Honey badger (subfamily Mellivorinae)
One species, *Mellivora capensis*.

A small brown blur streaking across a road; a striped snout peering cautiously out of a dark hole at dusk; a widening V-shaped ripple speeding away across still water; a rare glimpse of a graceful, brown cat-like creature in a tree: these are all that most people will ever see of a wild mustelid. Yet some of these shy animals (weasels, badgers, mink, skunks) are surprisingly common in north temperate farmland. Mustelids are also common in Africa and South America.

The mustelids are a large, widely distributed and rather mixed group. They occupy nearly every habitat, including fresh and salt water, in all continents except Australasia (though they have been introduced into New Zealand) and Antarctica. Many are small, under 2.2lb (1kg)—the smallest carnivores are mustelids—have a long body with short legs, and are skillful climbers. All have five toes on each foot, with sharp, nonretractile claws. Males are larger than females, particularly in weasels and polecats, where male skulls are some 5–25 percent longer, and body weights up to 120 percent greater (see p103). Sexual dimorphism is much less pronounced in badgers, otters and skunks. Mustelids have 28–38 teeth (see OPPOSITE).

As well as terrestrial weasels and polecats, the family includes the semiarboreal martens, amphibious otters, semiaquatic mink and burrowing badgers. The range of body weights is exceptional—the Giant otter and the wolverine may outweigh the Least weasel a thousand times. The basic short-leg, long-body plan also occurs with a variety of adaptations in form and diet fitted to life as a carnivore in very different habitats.

The anal glands are an important feature of the anatomy of most mustelids. They consist of two groups of modified skin glands, each emptying into a storage sac, which opens by a sphincter into the rectum near the anus. Discharge of the sacs is under voluntary control. The glands produce a thick, oily, yellow, powerful-smelling fluid called musk, the chemical composition of which is probably slightly different in each individual. A little musk is secreted with the feces, which are then carefully placed where other individuals can find them. Pine martens and sable often deposit them on conspicuous stones in the middle of a track; otters leave their spraints (feces) on the same riverbank sites for generation after generation. A secondary function of the glands is defense. When severely frightened most mustelids will discharge musk, probably as a reflex action. Perhaps from such beginnings, the musk glands of the New World skunks and some of the Old World polecats evolved into effective defense weapons, supported by unmistakable warning displays in their behavior and striking color patterns in their coats.

The reproductive habits of mustelids are remarkable for several unusual features. In most species the sexes live separately for much of the year; they rarely meet and are hostile when they do. During the temporary truce in the mating season the male seizes the female by the scruff of the neck and may drag her about vigorously before mounting. Copulation is repeated and very prolonged—up to one to two hours even in the weasels. The penis is stiffened by a bone, the baculum, which facilitates the long copulation. The whole procedure seems calculated to thoroughly arouse the female and is associated with induced ovulation. So far as is known, all female mustelids can be

▼ **Representative species,** illustrating the great variety of habitat and prey. (1) American mink with rabbit. (2) European polecat hunting in rabbit burrow. (3) Eurasian badger in tunnel of its sett. (4) Wolverine following scent trail across the ground. (5) Pine marten hunting birds. (6) Spotted skunk in threat posture which precedes spraying. (7) European weasel dragging mouse along a snow tunnel. (8) Cape clawless otter, using forepaws to hold down fish. (9) Pacific Sea otter about to crack shell of crustacean prey on stone lying on its chest.

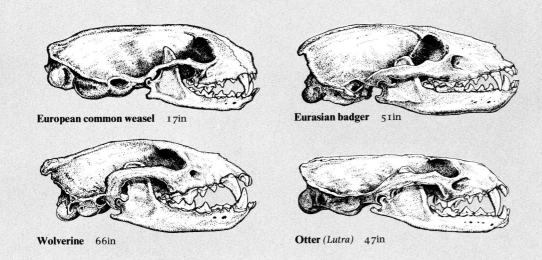

European common weasel 17in

Eurasian badger 51in

Wolverine 66in

Otter *(Lutra)* 47in

Skulls of Mustelids

Mustelid skulls tend to be long, flattened, and more or less triangular or wedge-shaped, tapering to the muzzle. Skull size, and numbers and adaptations of teeth vary widely. Most members of the largest subfamily, the Mustelinae, for example the European common weasel, have a dental formula I3/3, C1/1, P3/3, M1/2 = 34, with prominent, sharp canines and cutting carnassial teeth; in the wolverine (I3/3, C1/1, P4/4, M1/2 = 36) the heavy premolars and powerful jaws can crush even thick bones. The dental formula of the Honey badger is I3/3, C1/1, P3/3, M1/1 = 32, while in the Eurasian badger it is I3/3, C1/1, P4/4, M1/2 = 38—the largest number of teeth in the family. In otters of the successful genus *Lutra* the arrangement is I3/3, C1/1, P3–4/3, M1/2 = 34–36.

induced to produce eggs (ovulate) only by vigorous copulation. Prolonged mating may expose the pair to dangers from larger predators, but this risk is outweighed by the advantage of virtually certain fertilization.

After mating, the fertilized egg travels to the uterus, as is usual in mammals, developing, as it goes, into a ball of cells called a blastocyst. In most mammals, the blastocyst implants into the uterus wall within a few days and development of the embryo proceeds. But in the 16 or more mustelid species with delayed implantation, it floats free in the uterus for periods from a few days up to 10 months, and implants only when certain conditions are met. These are not the same for all species; for example, in New Zealand ermine implantation occurs once the lengthening days of spring reach a ratio of about 11 light to 13 dark hours. In Eurasian badgers, however, implantation occurs while they are semidormant in December. (See illustration on p13.)

Two puzzling aspects of delayed implantation in mustelids are its uneven distribution and its evolutionary advantage. So far as is known, extended delay occurs in all marten species and in the wolverine, but in none of the polecats; in the Eurasian and American badgers, but not in the Honey badger; in the North American river otter, but not in the closely related Eurasian river otter; in western forms of the Eastern spotted skunk, but not in eastern forms, nor in the Striped skunk; in ermine and Long-tailed weasels, but not in the Common weasel. There is no explanation as to why delayed implantation should benefit some species of river otters, spotted skunks and weasels but not similar species in the same or comparable habitats.   CMK

WEASELS AND POLECATS

Twenty-one species in 7 genera
(including 14 of 16 species of *Mustela*)
Family: Mustelidae.
Subfamily: Mustelinae.
Distribution: widespread from tropics to Arctic, in Americas, Eurasia, Africa; introduced in New Zealand.

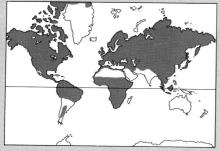

Habitat: very varied, from forests to mountains, farmland, semidesert, steppe and tundra.

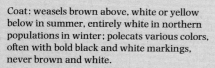

Size: ranges from head-body length 6–8in (15–20cm), tail length 1–1.5in (3–4cm) and weight 1–2.5oz (30–70g) in males of the Least weasel, to head-body length 18–22in (47–55cm), tail length 6.5in (16cm) and weight 3–7lb (1.4–3.2kg) in the grison.

Coat: weasels brown above, white or yellow below in summer, entirely white in northern populations in winter; polecats various colors, often with bold black and white markings, never brown and white.

Gestation: 35–45 days, extended by delayed implantation in 2 species.

Longevity: average less than 1 year in European common weasel, perhaps more in other species.

Species include: the **European common weasel** (*Mustela nivalis nivalis*), Eurasia and **Least weasel** (*M. n. rixosa*), N America. **European polecat** (*M. putorius putorius*), Europe and **ferret** (*M. p. furo*). **Ermine** or **stoat** (*M. erminea*), N America and Eurasia. **Long-tailed weasel** (*M. frenata*), N and S America. **Steppe polecat** (*M. eversmanni*), USSR to China. **Black-footed ferret** [E] (*M. nigripes*), N America. **Marbled polecat** (*Vormela peregusna*), Eurasia. **Zorilla** (*Ictonyx striatus*), Africa.

[E] Endangered.

▶ **"Giant" among predators**, a European polecat with its rabbit prey. Though smallest among the carnivores, members of the weasel and polecat group differ from most other members of the order in their ability to catch and kill, single-handed, prey much larger than themselves.

The Least weasel has been described as "the Nemesis of Nature's little people." The identification with the goddess of retribution is very apt for this formidable hunter, the smallest of a group of carnivores of different sizes but similar shape and habits. The seven genera of weasels and polecats comprise 21 of the 33 species of the subfamily Mustelinae.

Most weasels and polecats weigh less than 4.4lb (2kg), some much less, and there are pronounced differences in size between males and females (see RIGHT). Knowledge of the group is very uneven; the Northern Hemisphere (Holarctic) weasels—the European common weasel, the ermine or stoat, and the Long-tailed weasel—are known in much greater detail than any other species; for some of the tropical species there is no reliable information at all. As a group, these are among the most widespread carnivores in the world.

The Holarctic weasels have long, slim bodies and short legs, a flat-topped, sharp-faced, almost triangular head, and short, rounded ears. Coat color (see RIGHT) distinguishes them from the polecats, which in general are stockier in build and less agile, but have a similarly shaped face and ears. All species have the habit of sitting up on their haunches for a better view than they can get from on all fours.

All weasels and polecats are terrestrial hunters, taking whatever small rodents, rabbits, birds, insects, lizards and frogs are locally available. Most are purely carnivorous; their teeth are highly adapted to killing and cutting up prey (see p101). They do not feed indiscriminately—each species chooses, from the range of potential prey, a different menu. The choice is determined by, among other things, their own size and the size, relative number and catchability of possible food items. The small Common and Least weasels are hunters of mice and voles, especially Field voles; they can kill young rabbits, but it is probably not worth their while to try if they can still find enough small rodents and birds. The larger weasels, such as the ermine and Long-tailed weasel, catch any rodents that expose themselves above cover, but gain a better return in energy from concentrating on larger prey such as rabbits and water voles.

A natural experiment in Britain, started in 1953, illustrated the dependence of the Common weasel and ermine on different prey. The virulent introduced disease myxomatosis suddenly removed about 90 percent of the previously abundant rabbits. Hedges

COAT COLOR IN ERMINE

◀ **An ermine or stoat in summer coat,** chestnut colored with a white bib.

▼ **A transitional stage in the fall,** with white extending upward.

▽ **All-white winter coloration,** excepted only the black tip to the tail, not seen here.

The ermine is an opportunistic feeder, as these pictures demonstrate. Rabbits and voles may be the usual prey, but birds (here jay and Wood pigeon) and, near human habitation, domestic fowl eggs will eagerly be seized upon.

Why Are Male Weasels Bigger than Females?

Male mustelids are often larger than females. This is specially evident in the smaller species, such as the European common weasel (see RIGHT), where the female is half the weight of the male. The reason for this remarkable sexual dimorphism has perplexed scientists for generations, and a number of theories have been proposed.

The simplest explanation is that the size difference enables males and females to eat different prey and so avoid competition with each other, particularly when food is scarce. An individual will always try to prey on the largest animal possible, to get the best return for its effort; males are able to catch prey too big for females to handle.

An alternative, but not incompatible, explanation reflects the adaptation of males and females to quite different roles in reproduction. Males are promiscuous and territorial—they compete with other males for access to females and for the best territories. For them, fighting ability, actual or threatened, increases with size—the largest is often the most successful breeder and dominant individual, so sexual selection has favored the development of large males. For a female, who raises young on her own, small size has benefits; she can enter rodent

burrows inaccessible to other, larger predators (including male weasels), needs to eat less of her catch herself, and therefore need not travel so far from her young in the search for food.

and banks became flushed with vegetation, a home for many voles and mice. Throughout Britain the ermine population was drastically reduced, while that of the European common weasel increased. Since 1970 the slow recovery of the rabbits has accelerated and many field edges are close-cropped again. The ermine is now increasing, while the weasel declines.

Most of the larger polecats are more generalized feeders, eating anything they meet and can catch, most often rodents, but also insects, worms and carrion. One, the rare Black-footed ferret of the western prairies of the USA, is a true specialist, depending entirely on Prairie dogs. Where Prairie dogs are controlled by man, as they are over most of their range, Black-footed ferrets have died out. This is the only member of the weasel–polecat group listed as an endangered species by the IUCN.

The breeding habits are known in detail for only a few species. In the ermine and the Long-tailed weasel, implantation of the fertilized egg is delayed, but in other species apparently not. The young are born blind and thinly furred, in a secure nest, often borrowed from a prey species and lined with fur from previous meals. The eyes open at 3–4 weeks in the European common weasel,

5–6 weeks in the larger weasels and polecats. The young chew on meat at 3–5 weeks, although lactation lasts about 6–12 weeks and the family stays together some weeks after weaning. The larger species and males of the ermine and Long-tailed weasels first breed as one-year-olds; females of these species mate, when still nestlings, with the male currently holding the territory in which their mother resides. Both sexes of the European common and Least weasel may mature at 3–4 months if food is abundant.

Weasels hunt largely underground or under snow, and are active day and night. Some larger species are more often nocturnal. Usually, males and females have separate home ranges; these may overlap with the range of a member of the opposite sex, but not with the range of one of the same sex. Residents avoid each other whenever possible; females are subordinate to males, except when with young. Home ranges are smaller in habitats rich with prey, in the nonbreeding season, and in the smallest species. Male European common weasels in Britain have ranges of 2.5–62 acres (1–25 hectares); male ermine in Europe have ranges of 50–100 acres (20–40 hectares) and in Russia up to 250 acres (100 hectares); European polecats in Russia range over 250–6,200 acres (100–2,500 hectares). Grisons, and possibly the African striped weasel, may be more sociable, as they are reputedly seen in groups (of unknown composition) and, in captivity, one male and one female or two male grisons may be kept in one cage for long periods. However, four grisons observed together in captivity showed no evidence of the mutual grooming characteristic of social animals. Most mustelids, sociable or not, may be seen in parties of females and young just before the dispersal of the litters.

Some people persecute small mustelids as pests of game and poultry. The extinction of

◀ **Variations on a theme.** These chiefly more southerly species share the same body plan as the ermine or Common weasel but tend to have black, not brown as the predominant dark coloration, or to be larger. (**1**) North African banded weasel (*Poecilictis libyca*). (**2**) African striped weasel (*Poecilogale albinucha*). (**3**) Marbled polecat (*Vormela peregusna*). (**4**) The skunk-like zorilla (*Ictonyx striatus*), which appears to threaten to stink-spray. (**5**) Little grison (*Galictis cuja*) and (**6**) European polecat (*Mustela putorius*) in winter coats, both in upright sniffing/looking-out stance. (**7**) Patagonian weasel (*Lyncodon patagonicus*) in typical flattened weasel posture. (**8**) Black-footed ferret (*Mustela nigripes*) at Prairie dog burrow.

◀ **Gamekeeper's gibbet.** A dozen ermine (stoats) bear witness to the ruthless extermination of small mustelids as threats to game birds and poultry.

▼ **Trapper and marten** (see box).

Mustelids as Furbearers

Mustelids are the commonest of the small predators of the northern forests. In the vast snowy regions of Canada, Scandinavia and Siberia, sable and other martens, mink, kolinsky and ermine, otters and wolverines are active throughout the winter. These species have solved the problem of conserving the heat of their small bodies during the months of sub-zero temperatures by developing long, dense, water-repellent coats, which man greatly prizes.

Wild mustelids contributed substantially to the vigorous fur trade of the 18th and 19th centuries. At that time, wild-caught furs were important to the economy of northern lands. In the 16th to early 19th centuries, fur trappers and traders were among the first to explore and develop newly discovered North America. The furs of many wild mustelids were much sought after for their beauty and practical value. Furs such as Russian sable became a badge of wealth and rank; mink was, and still is, a byword for luxury; wolverine was prized as a trimming for parka hoods, because rime frost does not condense on it. Ermine was traditionally worn by British justices and peers—50,000 ermine pelts were sent from Canada for George VI's coronation in 1937—but nowadays the price of labor is so high that the tiny ermine pelts (300 for a coat) are not considered worth handling, especially as larger, equally fine, white pelts can be taken from other sources.

When the exploitation of furbearers was regulated only by an apparently insatiable market, the rapid price rises during the 19th century were bound to be followed by overtrapping. Populations of the larger and slower-breeding mustelids such as sable and fisher were greatly reduced by 1900. The possible disappearance of these economically important species stimulated much ecological research, especially in the USSR, where now (as elsewhere) fur trapping is carefully controlled by closed seasons and quotas adjusted annually to the estimated population densities of furbearers. (BELOW Trapper and marten.)

Some furs are now produced largely or entirely on farms, for example, American mink. The advent of acceptable synthetic fur fabrics, and changes in fashions, have reduced the demand for pelts. These developments ensure that what remains of man's exploitation of the wild furbearing mustelids is now more rational and sustainable. An exception is the Giant otter, still illegally poached in Brazil (see p117).

the European polecat in England was probably due to intensive gamekeeping. Others hail weasels and polecats as useful exterminators of rodents. Both opinions are exaggerated.

All weasels and polecats eat large numbers of voles, mice, rats and rabbits—in one year a single family may consume thousands of such prey. Farmers have always hoped (and as often assumed) that small mustelids could help rid their houses and farms of rodents—or at least prevent outbreaks. This was the attitude in Europe before the introduction of the domestic cat in the 9th century. It was still regarded as evident truth in 1884, when the European common weasel, ermine and ferret were deliberately introduced to New Zealand to control the European rabbits over-running the new sheep pastures. Unfortunately, they did not succeed, and neither did the Small Indian mongooses taken to Hawaii to clear sugar plantations of rats (see p140).

Small herbivores such as rabbits and rodents normally reproduce in greater numbers than their predators. If conditions become favorable for the rodents, they will increase rapidly, and the mustelids cannot reproduce fast enough to catch up. The mustelids will increase, but only after some delay, by which time the rodents are already abundant. Predation is usually heaviest when rodent numbers are declining for some other reason (for example, the increased age of maturity, shorter breeding season and more extensive dispersal which is characteristic of peak-year vole populations). In such circumstances mustelids can accelerate, even prolong, a decrease in rodent numbers.

In the extreme case of a small, isolated prey population with no safe refuges, mustelids can achieve impressive results. On an island off Holland, a few ermine introduced in 1931 increased rapidly and by 1937 had exterminated a plague of water voles. In a 21 acre (8.5 hectare) enclosure on a New Zealand farm, ferrets and feral Domestic cats together almost eliminated an entire dense population of rabbits in three years (up to 48 per acre or 120 per hectare). But out in the open fields, although mustelids may influence the way populations of voles fluctuate, they cannot "control" them, either by greatly reducing their numbers or by preventing new outbreaks. CMK

THE 21 SPECIES OF WEASELS AND POLECATS

Five species are sufficiently distinct to be placed in separate genera, and there are just two species of grison (*Galictis*). All others are *Mustela*, although some authorities recognize fewer species. Two distinct forms each of *Mustela nivalis* and *M. putorius* are here listed separately. For **European mink** (*M. lutreola*) and **American mink** (*M. vison*) see p108. Figures for size and breeding are mostly very approximate, or unknown (indicated by ?). Males of most species are considerably heavier than females (see p103).

European common weasel
Mustela nivalis

Europe from Atlantic seaboard (except Ireland), including Azores, Mediterranean islands, N Africa and Egypt, E across Asia N of Himalayas; introduced in New Zealand. Very large variation in size, from small form similar to Least weasel (see below) in N Scandinavia and N USSR—in Sweden HBL (male) 7–8in, TL 1–2in, wt 1–3oz—to largest in S beyond range of ermines—eg Turkmenia: HBL (male) 9–10in; TL 2–3in; wt to 9oz. Coat brown above, white below, turning entirely white in winter except in W Europe and S USSR; no black tip to tail. Gestation 34–37 days; Litter size 4–8; may produce 2, even 3, litters a year in vole plagues. (In Sweden two subspecies are recognized: *M. n. nivalis* in N and C Sweden is smaller, shows less sexual dimorphism, normally has white winter fur (but not always) and has more white on underside when compared to *M. n. vulgaris* in S Sweden, which retains summer coat in winter and also has a brown spot on cheek.)

Least weasel
Mustela nivalis rixosa

N America, S to about 40°N. HBL (male) 6–8in; TL 1–2in; wt 1–2oz. Coat brown and white in summer, turning white in winter. Breeding as for European common weasel. (Some scientists regard the Least weasel as a separate species, distinct from the European common weasel, even though they have interbred in captivity. Here it is considered a subspecies of *M. nivalis*.)

Ermine
Mustela erminea
Ermine, stoat or Short-tailed weasel.

Tundra and forest zones of N America and Eurasia, S to about 40°N, including Ireland and Japan, but not Mediterranean region, the semideserts of USSR and Mongolia, or N Africa; introduced in New Zealand. Very large variation in body size especially in N America; largest in N—HBL (male) 9in; wt 7oz—smallest in Colorado (where Least weasel is absent)—HBL (male) 7in wt 2oz; Russian races 4–7oz; British and New Zealand races up to 12oz. Coat brown and white; prominent black tip to tail, even in winter white. Delayed implantation of 9–10 months, from early summer mating to whelping in spring, which is not shortened by abundance of food; active gestation about 28 days; litter size 4–9, sometimes up to 18.

Long-tailed weasel
Mustela frenata

N America from about 50°N to Panama, extending through northern S America along Andes to Bolivia. HBL (male) 9–14in; TL 5–9in; wt (male) 7–12oz, (female) 3–7oz. Coat and breeding as in *M. erminea*; S races have white facial markings and yellow underparts.

Tropical weasel
Mustela africana

E Peru, Brazil. HBL (male) 12–13in; TL 8–9in; wt ? Coat reddish-brown above, lighter below with median abdominal brown stripe; black tail tip indistinct; foot soles naked. Formerly placed by some in separate genus *Grammogale*.

Mustela felipei

A new species first described in 1978, from highlands of Colombia. HBL of 2 males 8–9in; TL 4–5in; wt ? Coat blackish-brown above, orange-buff below; no black tail tip; short ventral brown patch (not stripe); feet bare and webbed. Formerly placed by some in separate genus *Grammogale*.

European polecat
Mustela putorius putorius

Forest zones of Europe, except most of Scandinavia, to Urals. HBL (male) 14–20in; TL 4–7in; wt 1–3lb. Coat buff to black with dark mask across eyes. Gestation 40–43 days; litter 5–8.

Ferret
Mustela putorius furo

Domesticated form of *M. putorius* (or possibly of *M. eversmanni*). Albinoes and white or pale fur common. Introduced and feral in New Zealand.

Steppe polecat
Mustela eversmanni

Steppes and semideserts of USSR and Mongolia to China. HBL (male) 12–22in; TL 3–7in; wt about 4lb. Coat reddish-brown, darker below and on feet and face mask; ears and lips white. Gestation 36–42 days; litter 3–6, occasionally up to 18.

Black-footed ferret E
Mustela nigripes

W prairies of N America, only within range of Prairie dogs (*Cynomys*); rare and endangered; derived from *M. eversmanni* invading N America during Pleistocene (2 million to 10,000 years ago). HBL (male) 15–16in; TL 4–5in; wt 2lb. Coat yellowish with dark facial mask, tail tip and feet. Gestation of one female in 2 seasons 42 and 45 days; litter 3–4.

Mountain weasel
Mustela altaica

Forested mountains of Asia from Altai to Korea and Tibet. HBL (male) 9–11in; TL 4–6in; wt 12oz. Coat dark yellowish to ruddy-brown, with creamy-white throat and ventral patches; paler in winter, but not white; white upper lips and chin shading to adjacent darker areas (cf *M. sibirica*). Gestation 30–49 days; litter usually 1–2 but up to 7–8.

Kolinsky
Mustela sibirica
Kolinsky, Siberian weasel.

European Russia to E Siberia, Korea and China, Japan and Taiwan (see also *M. lutreolina*). HBL (male) 11–15in; TL 6–8in; wt 23–29oz. Coat dark brown, paler below: may have white throat patch; paler in winter, but not white; dark facial mask with white upper lips and chin sharply contrasting with surrounding darker fur; tail thick and bushy. Gestation 35–42 days; litter 4–10.

Yellow-bellied weasel
Mustela kathiah

Himalayas, W and S China, N Burma. HBL (male) 9–11in; TL 6–7in; wt ? Coat deep chocolate-brown (rusty-brown in winter), yellow below; may have white spots on forepaws and whitish throat patch, chin and upper lips; tail long-haired, at least in winter.

Back-striped weasel
Mustela strigidorsa

Nepal, E through N Burma to Indochinese Peninsula. HBL (female) about 11in; TL 6in; wt ? Coat deep chocolate-brown (paler in winter), with silvery dorsal streak from base of skull to tail root and yellowish streak from chest along abdomen; upper lip, chin and throat whitish to ocherous; tail bushy; feet naked at all seasons.

Barefoot weasel
Mustela nudipes

SE Asia, Sumatra, Borneo. Coat uniform bright red with white head; feet naked at all seasons.

Indonesian mountain weasel
Mustela lutreolina

High altitudes of Java and Sumatra. The few known specimens are similar in size and color to European mink (*M. lutreola*) (russet-brown, no face mask, variable white throat patch); but skull similar to *M. sibirica*. Probably derived from *M. sibirica* stranded on the islands at the end of the Pleistocene. Some authorities consider *M. lutreolina* and *M. sibirica* as one species.

Marbled polecat
Vormela peregusna

Steppe and semidesert zones from SE Europe (Rumania) E to W China, to Palestine and Baluchistan. HBL (male and female) 13–14in; TL 5–9in; wt about 25oz. Coat black, marked with white or yellowish spots and stripes, face like European polecat, *M. putorius*. Gestation 56–63 days; litter 4–8.

Zorilla
Ictonyx striatus
Zorilla, African polecat.

Semiarid regions throughout Africa S of Sahara. HBL (male and female) 11–15in; TL 8–12in; wt 3lb. Coat black, strikingly marked with white; hair long and tail bushy. Gestation 36 days; litter 2–3.

North African banded weasel
Poecilictis libyca

Semidesert fringing the Sahara from Morocco and Egypt to N Nigeria and Sudan; closely related to *Ictonyx*, possibly same genus. HBL (male and female) 9–11in; TL 5–7in; wt (male) 7–9oz. Coat black, marked with variable pattern of bands and spots. Gestation unknown; litter 1–3.

African striped weasel
Poecilogale albinucha

Africa S of Sahara. HBL (male and female) 10–14in; TL 6–9in; wt 8–12oz. Coat black with 4 white and 3 black stripes down back; tail white. Gestation 31–33 days; litter 1–3.

Grison
Galictis vittata
Grison or huron.

C and S America from Mexico to Brazil, up to 3,940ft. HBL (male and female) 18–22in; TL 6in; wt 3–7lb. Face, legs and underparts black; back and tail smoky-gray with white stripe across forehead; feet partly webbed. Gestation unknown; litter probably 2–4.

Little grison
Galictis cuja

C and S America, at higher altitudes than *G. vittata*. HBL (male and female) 16–18in; TL 6–7in; wt about 2lb. Coat as *G. vittata*, but back is yellowish-gray or brownish.

Patagonian weasel
Lyncodon patagonicus

Pampas of Argentina and Chile. HBL (male and female) 12–14in; TL 2–4in; wt ? Top of head creamy-white; back grayish; underparts brown. Only 28 teeth.

▲ **European polecat.** The blond head and white feet of this animal indicate that it probably is a hybrid between a wild polecat and a domesticated ferret.

MINK

Two of 16 species of the genus *Mustela*
Family: Mustelidae.
Subfamily: Mustelinae.
Distribution: N America, France, E Europe to
NW Asia

Habitat: margins of waterways and lakes, rocky
coasts.

Gestation: 34–70 days.

Longevity: up to 6 or more years (to 12 in
captivity).

American mink
Mustela vison
American or Eastern mink.

Size: head-body length of males
13.5–21.5in (34–54cm), females
12–18in (30–45cm); tail length of
males 6–8.5in (15–21cm), of
females 5.5–8in (14–20cm).
Weight: 1.1–3.3lb (0.5–1.5kg).
Coat: thick, glossy, brown; winter
coat darker; white patch usually
lacking on upper lip. Fourteen
subspecies.

European mink
Mustela lutreola
Size: slightly smaller than American mink—
head-body length of males 11–17in
(28–43cm), females 12.5–15.5in (32–40cm);
tail length in males 4.7–7.5in (12–19cm),
females 5.1–7in (13–18cm). Weighs slightly
less than American mink. Coat similar to that
of American mink but with white patch on
upper lip. Seven subspecies, decreasing in size
from *M. l. turovi* (Caucasus) to *M. l. lutreola*
(northernmost form).

▶ **An American mink emerges** from its
waterside den. Mink are opportunistic hunters
and cache surplus prey in their dens. One
mink's cache was found to contain 13 freshly
killed muskrats, 2 mallard ducks and 1 coot.

A mink, to most people, is an expensive fur
coat. In reality, mink are two species of
lively carnivores which might be said to
cloak the shoulders of the Northern
Hemisphere, only the less fortunate ending
their days in the wardrobes of society ladies.

The American mink and the smaller and
less common European mink live predatory
lives along the margins of waterways. As a
result of escapes from fur farms, the
American mink is now naturalized in many
parts of Europe and has in places supplanted
the native species.

These close relatives of the weasels and
polecats are semiaquatic and have partly
webbed feet which assist in underwater
hunting. Mink are somewhat serpentine in
shape. They have small ears and long bushy
tails. The coat provides insulation against
low northern temperatures. The fur has two
components: long guard hairs each sur-
rounded by 9–24 underfur hairs that are
one-third or half the length. There are two
molts each year: the thick, dark winter coat
is shed in April, to be replaced by a much
flatter and browner summer coat. The summer molt occurs in August or September
and the winter coat is in its prime condition
by late November. Northern subspecies
have darker fur than southern forms.

Mink originally evolved in North
America. The European species is a late
migrant to Eurasia across the Bering Land
Bridge during the last glacial phase of the
Pleistocene. The two species have only been
geographically isolated for some 10,000
years and are, in consequence, very similar
in appearance, although there are skeletal
differences, and American mink grow to a
greater size (as do the males of each species).

Mink are truly carnivorous and take a
wide variety of prey from aquatic and bank-
side habitats. Their eyesight is not particularly well adapted to underwater vision,
and fish are often located from above before
the mink dives in pursuit. Mink rely heavily
upon sense of smell when foraging for
terrestrial prey.

Mink are solitary and territorial. Individuals defend linear territories of 0.6–2.5mi
(1–4km) of river or lake shore by scent
marking and overt aggression. Each territory contains several waterside dens, and a
"core" area where the occupant forages
most intensively. Marshland territories
cover up to 22 acres (9 hectares), while
those on a rocky coastline, such as
Vancouver Island, are only 0.4mi (0.7km)
long, reflecting the replenishment of rock-
pool resources by tides. Female mink ranges

◄ **Distinguishing two species.** The European mink BELOW always has a white patch on its upper lip. American mink ABOVE, now naturalized in Europe, are larger, and most lack the white patch. However, 10–20 percent not only possess the patch, but it may be as large as that of the European species. In such animals only study of the skeleton can guarantee correct identification.

The Versatile Mink

The carnivorous diet of mink includes crayfish, crabs, fish (1), small burrowing mammals (2), muskrats and rabbits, and birds (3). This range of prey—hunted in water, on land in swamps and down burrows—is considerably greater than that of more specialized mustelids, such as otters and weasels.

For mink, the so-called "broad niche" which they occupy carries both costs and benefits. The costs arise when mink compete with a more specialized predator. Since a "specialist" is better adapted than a "generalist" to exploiting certain prey, the generalist fares the worse when those prey become the object of competition. For example, at times of absence or scarcity of other prey groups, mink may depend heavily on fish, bringing them into direct competition with otters. Such competition only limits mink populations when otter population density is high. Even then direct competition is normally restricted by other factors, such as differential selection of fish prey on a size basis and exploitation of different parts of the habitat (for example, on a lake, the open waters by otters, and marsh and reedbeds by mink). On the benefit side, mink have such a wide choice of prey that they can normally turn to alternatives if one type of prey becomes scarce; such an option is closed to specialists.

In other respects mink are also at an advantage in a waterside habitat. Their small size, allowing access to many diverse refuges and better use of available cover, and tolerance of human disturbance give them the edge on, for example, otters, which are intolerant of humans and require dense riverside cover and larger holt sites.

The adaptability of minks is reflected in the variety of habitats in which they thrive—from the arctic wastes of Alaska, to the steaming swamps of the Florida Keys; from inland lakes and rivers to wave-battered rocks of the Atlantic coastline.

are about 20 percent smaller than those of males quoted above. Mink scent mark with feces coated by secretion from glands at the end of the gut (proctodeal glands), by an "anal drag" action, and by secretions from glandular patches on the skin of the throat underside and chest, deposited by "ventral rubbing."

As the mating season (February to March) approaches, males leave their territories and travel long distances in search of females. One male may mate with several females and each female may be mated by several males. How, with promiscuous mating, do the fittest individuals contribute the most offspring to the next generation? It appears that the roving existence of the rutting male is very demanding, thus ensuring that stronger animals travel farther and mate with more females than weaker ones. Experiments in mink farms have shown that when a female is mated by several males during her three weeks on heat, it is the last mating which produces most of the kits. In the wild, therefore, the males which father the most kits are the stronger ones which are still mating at the end of the season. Fighting is common between rutting males.

Seven to 30 days may elapse between fertilization and implantation of the egg; gestation proper lasts 27–33 days. After the resulting five- to ten-week pregnancy four to six blind and naked young are born. The female rears the kits alone and weans them at 8–10 weeks. They disperse from her territory at 3–4 months of age, males to a greater distance, often 31mi (50km) or more. Sexual maturity is reached at 10 months.

The American mink is widely regarded as a pest and a possible threat to native species in countries where it is now naturalized. The European mink may be one of those threatened species; already declining as a result of intensive hunting, it may fail in competition with its more vigorous American relative. Although it is protected in some countries, the European mink's conservation may be further hampered by problems of identification and hybridization where the two species occur together.

Mink have been trapped for their fur for centuries. The American species has a superior coat and has been bred in fur farms since 1866. Its hardiness and variety of mutant fur colors make it especially suited to this purpose. In 1933 the Russians started a program of release into the wild in order to establish a superior source of "free range" fur; by 1948 3,700 had been released.   JDSB

MARTENS

All 8 species of the genus _Martes_
For tayra (_Eira barbata_) see opposite.
Family: Mustelidae.
Subfamily: Mustelinae.
Distribution: Asia, N America, Europe.

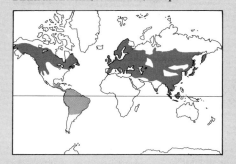

Martens   Tayra

Habitat: forests, chiefly coniferous but also deciduous and tropical mountain forests; Stone marten in urban areas.

Size: head-body length 12–30in (30–75cm); tail length 4.7–18in (12–45cm); weight 1–11lb (0.5–5kg). Males 30–100 percent heavier than females in all species.

Coat: generally soft, thick, brown, with feet and tail darker, sometimes black, and often a pale throat patch or bib; tail bushy; soles of feet furred.

Gestation: 8–9 months (including 6–7 month delay in egg implantation) in most N species.

Longevity: to 10–15 years in most species.

▲ **A Pine marten in its hollow-tree nest** with a captured mouse. Pine marten numbers have been much reduced in many parts of Europe. At the western limit of its range it crosses with sable, producing a hybrid called the "kida."

THE North American porcupine would seem to have the perfect defense against predators – sharp quills. But the fisher, one of eight marten species, has evolved a unique technique to kill porcupines and thus is the only animal for whom this porcupine is an important prey item. Foraging both in trees and on the ground, the fisher, like all martens, is a highly adapted, efficient predator.

Martens are medium-sized carnivores, only moderately elongated in shape, with wedge-shaped faces and rounded ears that are larger than in some mustelids. While the bushy tail serves as a balancing-rod, the large paws with haired soles and semi-retractile claws are also great assets to these semiarboreal animals. Martens seem to leap from branch to branch effortlessly, and are among the most agile and graceful of the weasel family. Pathways, in trees or on the ground, may be marked with scent from the anal glands and with urine.

The Stone, Beech or House marten inhabits coniferous and deciduous woodlands from Europe to Central Asia and is often found near human habitation. Its large white throat patch extends onto its forelegs and underside. It is very similar to those species restricted to northern coniferous forests—the Pine marten, sable, Japanese marten and North American marten, which have smaller bibs. Differences between these four (which some authorities classify as one species) are graded from Europe eastward to North America: for example, the Pine marten is the largest and the North American marten the smallest. Two species from southern Asia, the Yellow-throated marten and Nilgiri marten, have striking yellow bibs and are again sometimes considered one

species. The fisher, from coniferous and mixed woodlands of North America, is the largest species and lacks the throat patch of other martens.

Martens are opportunistic hunters whose main foods are small vertebrates, especially mice, squirrels, rabbits and grouse. Carrion is also important in their diet, as are fruits and nuts when abundant. Martens inspect likely prey hiding places and if they sight prey, attempt a short rush to catch and kill it with a bite to the back of the neck.

Fishers are typical martens in most respects. They are not named for skill at catching fish or "fishing" bait out of traps as is commonly supposed. More likely, the name derives from old English ("fiche"), Dutch and French words for the European polecat and its pelt. Fishers are best known for their unique technique for preying on porcupines. The arrangement of quills on a porcupine protects it from an attack to the

The Tayra—a South American Marten?

The tayra (*Eira barbata*) inhabits forests from Mexico to Argentina and in Trinidad, where it searches for birds, small mammals and fruits—sometimes causing substantial damage to banana crops.

Tayras are similar in size and form to fishers. Head-body length is 35–45in (90–115cm), tail length 14–18in (35–45cm) and weight 8.8–13lb (4–6kg). Their coat is shorter, dark brown to black in color, with the head gray, brown or black and a yellow to white throat patch. Litters average three kits and are born in May after a 63–67-day gestation without the delayed egg implantation that occurs in most martens.

The comparison of the tayra with the North American fisher is a common one. But a correlation of typically mustelid features (elongation in shape, difference in size between sexes, carnivorous diet, and intolerance of other members of the same species) shows that the fisher and other martens (and also the small weasels) are at one end of a spectrum, and the tayra (and perhaps the badgers) closer to the other.

Tayras are less elongate in shape, have longer legs and are less sexually dimorphic in body size than are the martens. Tayras are fairly tolerant of other members of their

species and are often found in pairs and larger, probably family, groups, while martens are intolerant of other members of their species and are territorial toward members of their own sex. Vegetable matter, especially fruits, is more important in the diet of the tayra than in that of martens.

Finally, the tayra has a metabolic rate that is lower than might be expected for a mammal of its size, while the fisher is known to have a slightly elevated metabolic rate.

No good explanations of these correlations have yet been presented, although rates of energy expenditure may be part of the explanation. Species that are more elongate and live in a cooler climate expend energy more rapidly and thus have higher energy requirements than does the more compact tayra with its tropical distribution.

◄ **The Beech marten primarily hunts in trees** ABOVE, but may hunt rats and mice around farms, even denning in attics, and also inhabits rocky areas—hence the alternative names House marten and Stone marten.

◄ **American marten foraging in winter snow.** This species prefers continuous coniferous forest. Although populations were decimated in the first half of the 20th century, the species is now making a comeback.

back of the neck, where most carnivores attack, but its face is not protected. Fishers stand low to the ground and can thus direct an attack to a porcupine's face, yet are big enough to inflict damaging wounds. A fisher circles the porcupine on the ground, taking advantage of any chance to bite its face. The porcupine attempts to keep its well-quilled back and tail toward the fisher and to seek protection for its face against a log or tree. If the fisher delivers enough solid bites to the porcupine's face, the porcupine suffers shock or is unable to protect itself. The fisher then overturns the porcupine and begins feeding on its unquilled belly.

Killing a porcupine is long, hard work and a successful kill may take over half an hour. Depending on how many scavengers share the kill, the fisher may have enough food for over two weeks. Where porcupines are common, they may make up a quarter of a fisher's diet.

Martens are generally solitary. They are polygamous, and mating normally occurs in late summer (early spring in the fisher). During early spring, litters of 1–5 sparsely furred, blind and deaf kits are born. Around 2 months of age the kits are weaned. They are able to kill prey by 3–4 months, shortly before they leave their mothers. Martens become sexually active when between one and two years old, depending on species and sex, and it is probably at this time that they try to establish territories.

Martens were a distinct group within the weasel family by the Pliocene (7–2 million years ago). Skeletal characteristics of these ancestral martens show that the three present-day groups of martens—Pine martens, Yellow-throated martens and the fisher—were already distinguishable. Evolution of the distinct species within the former two groups began during the Pleistocene only 2 million years ago.

Because several of the martens, especially the sable and fisher, have been valued for their fur, hunting and trapping pressure on marten populations has sometimes been very high. This pressure, in combination with the destruction of the conifer and conifer-hardwood forests preferred by these species, has led to a decline in some populations. At present, however, none of the martens is considered endangered. **RAP**

Abbreviations: HBL = head-body length; TL = tail length; wt = weight.

Pine marten
Martes martes

C and N Europe, W Asia. HBL 16–22in; TL 8–11in; wt 2–4lb. Coat chestnut-brown to dark brown, bib creamy-white.

American marten
Martes americana

Northern N America to Sierra Nevada and Rockies in Colorado and California. HBL 12–18in; TL 6–9in; wt 1–3lb. Coat golden-brown to dark brown, bib cream to orange.

Japanese marten
Martes melampus

Japan, Korea. HBL 12–18in; TL 6–9in; wt 1–3lb. Coat yellow-brown to dark brown, bib white/cream.

Fisher
Martes pennanti
Fisher or pekan or Virginian polecat.

Northern N America to California (Sierra Nevada) and W Virginia (Appalachians). HBL 18–29in; TL 12–16in; wt 4–11lb. Coat medium to dark brown; gold to silver hoariness on head and shoulders; legs and tail black; variable cream chest patch. Gestation 11–12 months (implantation delayed 9–10 months).

Sable
Martes zibellina

N Asia, N Japanese islands. HBL 14–22in; TL 5–7in; wt 1–4lb. Coat dark brown, yellowish bib not always clearly delineated; tail short. Egg implantation and gestation 1 month longer than above species.

Stone marten
Martes foina
Stone, Beech or House marten.

S and C Europe to Denmark and C Asia. HBL 17–22in; TL 9–12in; wt 1–4lb. Coat chocolate-brown, underfur lighter than previous species, white bib often in 2 parts.

Yellow-throated marten
Martes flavigula

SE Asia to Korea, Java, Sumatra, Borneo. HBL 19–27in; TL 14–18in; wt 2–11lb. Coat yellow-brown to dark brown, bib yellow to orange, legs and tail dark brown to black. Gestation variable, 5–6 months with some 3–4 months' delayed implantation.

Nilgiri marten
Martes gwatkinsi
Nilgiri or Yellow-throated marten.

Nilgiri mountains of S India. Smaller than *M. flavigula*, but coat similar.

WOLVERINE

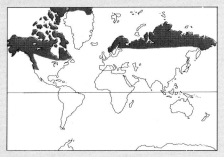

Gulo gulo
Wolverine or glutton.
Sole species of the genus.
Family: Mustelidae.
Subfamily: Mustelinae.
Distribution: circumpolar, in N America and Eurasia.

Habitat: arctic and subarctic tundra and taiga.

Size: in Alaska, head-body length up to 33in (83cm); tail length 8in (20cm); weight up to 55lb (25kg); in Alaska males average 33lb (15kg), females 22lb (10kg).

Coat: long, dark brown to black; lighter band along flanks to upperside of bushy tail.

Gestation: about 9 months.

Longevity: to 13 years (18 in captivity).

The 2 subspecies are the **European wolverine** (*Gulo gulo gulo*) and the **North American wolverine** (*Gulo gulo luscus*).

▶ **Over soft, deep snow** RIGHT, ABOVE the large feet of the wolverine enable it to catch its reindeer prey, which is handicapped by a weight load 8–10 times greater.

▶ **Somewhat bearlike in outward appearance,** the powerful wolverine is occasionally killed by packs of wolves (and probably also by Grizzly bears and pumas where they occur with wolverines). But a solitary wolf would find a wolverine a fearsome combatant.

▶ **The wolverine's remote habitat** has not protected it from persecution: it is hunted by fur trappers in North America and in Scandinavia. In consequnce, its range and numbers have decreased during the past 100 years, though some expansion of range is occurring where the wolverine is protected or its harvest regulated.

BECAUSE it is rare and inhabits remote areas, the wolverine has been poorly understood for centuries. The first description of the "glutton"—in 1518—tells of "an animal which feeds on carcasses and is highly ravenous. It eats until the stomach is tight as a drumskin then squeezes itself through a narrow passage between two trees. This empties the stomach of its contents and the wolverine can continue to eat until the carcass is completely consumed." This fable was still widespread during the 18th century. At that time even the famous Swedish taxonomist Carl Linnaeus was uncertain whether the European wolverine belonged to the weasel family or to the dog family; in the first edition of his *Systema naturae* (1735) the wolverine was even omitted.

The wolverine is heavily built, with short legs. However, because its feet are large they bear a weight load of only 0.4–0.5lb/sq in (27–35g/sq cm). In consequence, although an adult reindeer can elude a wolverine on bare ground, or even on snow if the crust is thick enough, the wolverine has the advantage in soft snow. In winter the wolverine feeds mainly on reindeer and caribou, which are either killed or scavenged. The wolverine's skill in scavenging indicates that it has a particularly well developed sense of smell. While it will kill a small mammal by a neck bite and usually eat it immediately, a wolverine drags down larger prey by jumping on its back and holding on with its powerful claws until the animal collapses to the ground. The wolverine's powerful jaw and chewing muscles enable it to break even thick bones. A carcass, although often completely utilized, is not immediately consumed by a wolverine but dismembered and hidden in widely dispersed caches, in crevices or buried in marshes or other soft ground. These provisions may be used, as much as six months later, by a wolverine female to feed herself and her newborn kits. The more diverse summer diet includes birds, small and medium-sized mammals, plants and the remains of reindeer calves or other prey killed by predators such as lynx, wolves and Grizzly bears.

The mating season extends from April to August, but implantation of the fertilized egg is delayed so that births take place during late January to early April, when 1–4 (average 2.5) kits are born blind, in a den that is usually dug in a deep snowdrift. The kits stay in and around the den until early May, but litters may be moved to new sites. Such moves are believed to be triggered by meltwater entering the den or by direct disturbance. Moves may also be an adaptation to avoid human persecution, since the accumulation of tracks around the den makes it easier for hunters to locate it. Females reach reproductive maturity in their second year, but reproduction is suppressed when food is scarce, whereas when prey is bountiful females breed annually.

The young are accompanied by their mother throughout the summer and thereafter remain, living within her home range at least into late fall. Male kits generally disperse by the onset of the next breeding season, but some female kits may remain in or near their mother's home range indefinitely. The adult female maintains, by scent marking and aggression, a territory of 19–135sq mi/(50–350sq km), which is exclusive from April to September. Male territories span up to 230–390sq mi (600–1,000sq km) and overlap the areas of several females as well as other males. Scent marking is usually in the form of urination, defecation or a scent secretion from the abdominal or ventral gland. In the latter case, a wolverine traveling over tussocky tundra will periodically stop while straddling a tussock and rub the scent gland on it. Willow shrubs may be similarly marked as individuals clamber over branches.

The distribution of wolverines has contracted during the past 100 years, for example from eastern Canada and the prairie provinces, and their numbers have also declined over large areas. These reductions are most marked where human populations are densest. The wolverine is not endangered, but it may be considered as vulnerable over most of its range.

Wolverines are shot and trapped where they are still regarded as big game, as a predatory pest or as furbearers, although the toll is modest (recently 800 pelts per year from Alaska, each with a value of $150–250) compared to that during the 16th and 17th centuries, when wolverine pelts were so valuable that the city of Turinsk in Siberia reputedly incorporated a wolverine in its coat of arms as the symbol of local commerce. AB/AJM

SKUNKS

Subfamily: Mephitinae
Thirteen species in 3 genera.
Family: Mustelidae.
Distribution: N, C and S America.

Size: overall length ranges from 16in (40cm) to 27in (68cm), weight from 1.1lb (0.5kg) to 6.6lb (3kg).

Longevity: to at least 7 years (to 8–10 years in captivity).

Skunk rabies
Red Fox rabies

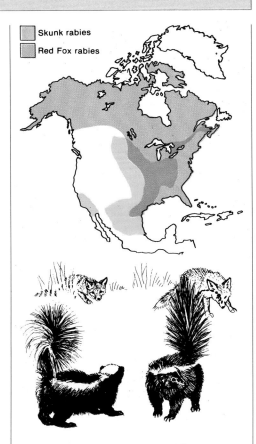

▲ **Skunks and rabies.** Over much of the USA skunks are the chief vector (carrier) of rabies. Where both skunks and Red foxes are carriers, rabies is often transmitted when a rabid fox attacks a skunk despite the threat display and squirting tactics ABOVE that usually deter a healthy predator.

▶ **The Pygmy spotted skunk,** a rare species of western and southwestern Mexico, showing the silkier coat that is characteristic of all Spotted skunks.

SKUNKS are best known for their odorous defense and their role as a transmitter (vector) of rabies. Although all mustelids possess anal glands, the ability to expel a fine spray of foul-smelling liquid at an intruder is most pronounced in skunks, in which it develops at less than one month old. Most species of skunk forewarn predators by stamping their front feet, raising their tail and walking stiff-legged. Spotted skunks will occasionally bluff by handstanding without spraying. If that fails, they will drop onto all fours and spray. The spray is aimed at the face and causes intense irritation, even temporary blindness, if it reaches the eyes. The odor (a sulfurous smell) persists in the area for days and absorbent clothing that has been "sprayed" is probably best discarded. Although most animals avoid skunks, a notable exception is the Great horned owl, which does not appear to be deterred by skunk spray while foraging at night.

Skunks are intermediate between weasels and badgers in build. There can be considerable variation in coat pattern even within species: for example in the Hooded skunk the white-backed form has two bands of white on its back narrowly separated by a black line, whereas in the black-backed form the white stripes are widely separated and are situated on the sides of the animal. The distinctive black and white markings help advertise their presence to intruders and, together with threat postures that make use of the tail covered with extra-long hairs, forewarn them that they may be sprayed. Skunks can spray up to 13–23ft (4–7m) in a favorable wind—although they are usually only accurate for up to about 2m (6.5ft). Normally skunks and skunk dens do not smell "skunky." During aggressive encounters between skunks, however, they will spray one another.

All skunks are largely carnivorous, with insects and small mammals as major prey, but they also eat grubs, birds' eggs and fruit seasonally. They are found in a wide variety of habitats, and are common in many urban areas, but prefer open or forest edge areas, where they forage at night, using their long front claws for rooting out food.

Inactive denning periods (not true hibernation) occur during the winter months if weather conditions are severe. In some areas several species occur together, but each uses different portions of the area more effectively than the others. This habitat partitioning allows coexistence, although the Striped skunk—the most common species—is invariably dominant where it overlaps with others. Except for Spotted

skunks, which are mainly active at twilight, skunks are chiefly nocturnal. For example, striped skunks normally forage only at dusk, dawn and during the night, ambling in search of prey at a leisurely pace and avoiding contact with people and domestic animals.

Skunks are the major vectors of rabies in much of the continental United States, with over 4,000 cases diagnosed in skunks during some years—in other words over two-thirds of all cases in wildlife. Whereas the role of the Red fox in rabies outbreaks has been declining recently, the role of skunks is increasing. The skunk's role in carrying rabies in some areas led early homesteaders to name it the "Phoby cat" or "Hydrophobia skunk." Rabid skunks may attack virtually anything that moves. All species carry rabies, but the Striped skunk is most often involved as it is more widespread. Skunk rabies cases are most prevalent in the midwestern USA and recently have been the most important source of human exposure to rabies in Texas and California. Rabid skunks have high levels of virus in the saliva, and as they have long incubation periods they are important reservoirs of the disease, which might otherwise be easier to control among foxes. Skunk spray is not known to carry rabies virus.

Skunks coexist with foxes, raccoons and coyotes, groups of skunks often using the same burrows as these species, but at different times of the year. Since skunks are in frequent contact with man's domestic animals, such as cattle, horses, pigs, dogs and cats, there is a great deal of potential for interspecies rabies transmission. Rabies outbreaks occur when skunks' movements are most extensive, during the fall and spring months. Transmission between skunks may occur during winter communal denning when territories overlap, and also through

▲ **Species of skunks.** (1) Western spotted skunk in a handstand threatening posture characteristic of the genus *Spilogale*. (2) Hooded skunk. (3) Hog-nosed skunk foraging with its long naked snout. (4, 5) Two forms of the Striped skunk, the commonest species.

the aggressive behavior of some males towards females with litters. Since rabies reduces population density and contact between individuals, outbreaks often occur 3–4 years apart when populations are high.

During most of the year females occupy home ranges of 0.4–0.8sq mi (1–2sq km),

each overlapping at least partially with other females. The territory of one male will encompass those of several females, but rarely that of other males. Males have no role in raising young. In fact, aggressive behavior by adult males toward females and their young can result in deaths.   DRV

Abbreviations: HTL = head-to-tail-tip length; wt = weight.
[*] CITES listed.

Striped skunk
Mephitis mephitis

S Canada, USA, N Mexico. Commonest species; diverse habitats include suburban areas; dens in burrows and even under buildings. Overall length 27in, weight 3–7lb. Coat black with forking white stripes of varying length on back and tail; white patch on head; black and white hairs not mixed. Mates in February; true gestation 62–66 days; 3–9 young born in May–June.

Hooded skunk
Mephitis macroura

SW USA. Rare, more secretive than

M. mephitis. Coat black with white back; 1–3 white stripes may be present; black and white hairs mixed. Mates in February; true gestation 63 days; 3–5 young born in May.

Hog-nosed skunk
Seven species of the genus *Conepatus*

Hog-nosed skunk (*C. mesoleucus*), S USA, Nicaragua; **Eastern hog-nosed skunk** (*C. leuconotus*), E Texas, E Mexico; **Amazonian skunk** (*C. semistriatus*), S Mexico, Peru, Brazil; **Andes skunk** (*C. rex*), Peru; *Conepatus castereus*, Argentina; *C. chinga*, Chile, Argentina, Paraguay; **Patagonian skunk** [*] (*C. humboldti*), Patagonia.

Diverse habitats, but prefer rugged terrain; den in rocky crevices and burrows. Overall length 60cm, weight 3–4lb. Coat black with large white band on back and white tail; color may be similar to *M. mephitis* where the two do not overlap; distinguished by lack of white head stripe and by bare elongated snout; large claws another adaptation for digging. Mate in February; true gestation 42 days; 2–4 young born in May–June.

Spotted skunks
Four species of the genus *Spilogale*

Western spotted skunk (*S. gracilis*), W USA to C Mexico. **Eastern spotted skunk** (*S. putorius*), SE and C USA to E

Mexico. **Southern spotted skunk** (*S. angustifrons*), C Mexico to Costa Rica. **Pygmy spotted skunk** (*P. pygmaea*), W and SW Mexico. All readily climb trees, den in crevices, burrow and use low buildings. All some 16in in overall length, weighing 1lb. Coat black with 4–6 broken white stripes or spots; hair silkier than in other genera. All bear 2–6 young. *S. angustifrons*, *S. putorius* and *S. pygmaea* mate February–March; true gestation 42 days; young born in May in *S. putorius* true gestation may be delayed 2 weeks. *S. gracilis* mates in late summer; with delayed implantation young are born in April–May.

OTTERS

Subfamily: Lutrinae [•]
Twelve species in 6 genera.
Family: Mustelidae.
Distribution: widespread in subpolar regions excluding Australasia and Madagascar, the Sahara, part of E USSR, W China, W Asia, S North America, SW South America.

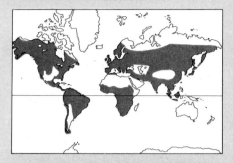

Habitat: aquatic (including marine), and terrestrial.

Size: ranges from the Oriental short-clawed otter with head-body length 16–25in (41–64cm), tail length 10–14in (25–35cm), weight about 11lb (5kg), to the Giant otter of S America with head-body length 38–48in (96–123cm), tail length 18–26in (45–65cm), weighing up to 66lb (30kg). The shorter Sea otter may attain 88lb (45kg).

Coat: browns and grays; darker above, chest, throat and underside usually lighter.

Gestation: mostly 60–70 days, but up to 12 months with delayed implantation in some species.

Longevity: up to 12 years in wild (Giant otter), and 21 years in captivity.

Six species of the genus *Lutra*: the **North American river otter** (*L. canadensis*), N America; **European river otter** or **Eurasian otter** (*L. lutra*), Eurasia, N Africa; **Marine otter** [v] (*L. felina*), S America; **Southern river otter** (*L. provocax*), S America; **Neotropical river otter** (*L. longicaudis*), C and S America. **Hairy-nosed otter** [I] (*L. sumatrana*), SE Asia.

Spot-necked otter (*Hydrictis maculicollis*), Africa.

Indian smooth-coated otter (*Lutrogale perspicillata*), Asia.

Oriental short-clawed otter (*Aonyx cinerea*), Asia; **Cape clawless otter** (*A. capensis*), Africa.

Giant otter [v] (*Pteronura brasiliensis*), S America.

Sea otter (*Enhydra lutris*), N America.

[•] CITES listed. [v] Vulnerable.

[I] Known to be threatened, status indeterminate.

ALTHOUGH all are rather similar in appearance, otters exhibit striking behavioral and social variations; in most species numbers are rapidly declining, through direct persecution by man, and indirectly because of loss of habitat.

Otters are the only truly amphibious members of the weasel family. They largely forage in water, but are equally at home on land, except for the Sea otter, which rarely comes ashore in California. Otters' tightly packed underfur and long guard hairs are water-repellent, and the body is elongated, sinuous and lithe, built for vigorous swimming. The limbs are short and the paws are webbed in most species. The forefeet are shorter than the hindfeet. The tail is fully haired, thick at the base and tapering to a point, flattened on the underside and in some species on the upper surface. There are numerous stiff whiskers (vibrissae) around the nose and snout, and in tufts on the elbows. These tactile hairs are sensitive to water turbulence and are used in searching for prey. The ears are small and round, and like the nostrils are closed under water, an indication that sound is not important in the location of prey. Most otters have claws, but those without them (and also the Sea otter) use their acute sense of touch and their manual dexterity to find and seize prey.

Otters' diets are varied, but most species eat frogs, crayfish, crabs and fish—usually sluggish non-game species such as roach, sticklebacks or eels. Otters consume their catch immediately. Bones are crushed with well-developed premolars (see p101 for dentition of otters). Very occasionally a rogue will enter a hatchery and kill large numbers of fish. Like most carnivores, otters prey on what is readily available and the easiest to catch: in Europe lethargic frogs and eels in winter, coots and ducklings in spring, spawning crayfish, perch or roach in summer.

Otters use different tactics to catch their prey; the Oriental short-clawed and Cape clawless otters (both *Aonyx* species) are hand-oriented and will invariably grab an octopus or a crayfish with outstretched forepaws; on the other hand, the six *Lutra* species and the Giant otter are mouth-oriented like most carnivores, so chase their fish prey underwater, catching it in their jaws. These successful predators can quickly hunt down and eat their daily quota of food, which may be about 2.2lb (1kg) for *Lutra*.

An otter will have several hunting sessions a day, swimming and feeding for an hour or more before hauling out to rest on the bank. Otters have a very rapid metabolism and a meal will pass through their digestive tract in a few hours—giving them boundless energy on the one hand, but forcing them also to eat at frequent intervals, often four times a day. The Giant otter and Indian smooth-coated otter, which often hunt in pairs or groups, remain in the vicinity of their still-feeding partners, waiting for them to finish before moving on.

▲ **The hydrodynamic form** of this Small-clawed otter exemplifies the perfect adaptation of all otters to an amphibious way of life.

▼ **Stiff whiskers and manual dexterity** are adaptations to catching prey in muddy or dark waters. Here a Cape clawless otter eats a fish.

The otters of the genus *Lutra* are probably the most numerous and are certainly the most widespread otters. The four New World species range from Alaska to Tierra del Fuego. The three Central and South American species are so similar in size and shape that they can only be distinguished by the shape of the nose pad (see p119). Four previously separated species (*L. annectens, L. platensis, L. enudris, L. incarum*) have recently been lumped together under the heading *L. longicaudis*; it is even suggested that all South American otters, while showing geographic variation in size and color, are really subspecies of the North American river otter or Canadian otter (*L. canadensis*). One New World species, the Marine otter, inhabits the rough waters of the western coast of South America from Peru to Cape Horn. Its fur is coarse and the skull is similar to that of the Spot-necked otter of Africa. Unlike the true Sea otter of the northern Pacific, this small otter's bones and teeth are not modified to cope with an essentially marine existence. Darwin, during the voyage of the *Beagle* (1831–36), reported that the natives of Tierra del Fuego ate it and used its warm fur for making hats.

The most widespread Old World representative of the genus is the European or Eurasian river otter, which ranges from Scotland to Kamchatka and south to Java. The Sumatran otter, last of the *Lutra* species, has a hairy nose like the Giant otter of Brazil, and is a spot-necked species. It lives in the high mountain streams of Southeast Asia.

The Spot-necked otter, one of the most proficient swimmers of all freshwater otters, inhabits streams and lakes of Africa south of the Sahara.

The Indian smooth-coated otter is more heavily built and larger than the Eurasian otter. It probably evolved earlier than the present-day *Lutra*. Like the Giant otter, it has short, dense fur, a flattened tail and thickly webbed paws which are nonetheless remarkably agile in manipulating and retrieving small objects. The shortened face and domed skull house broad molars, indicating a largely crustacean diet for this marsh-dwelling species; in Sumatra it even leads a coastal existence, earning it the confusing local name of "sea otter." Although sometimes included in the genus *Lutra*, it has quite distinct skeletal and behavioral characteristics.

The genus *Aonyx* contains two species, both of which are hand-oriented. The Oriental short-clawed otter is the smallest of all otters, rarely more than 35in (90cm) in overall length. Its forefeet are only partially webbed and have stubby agile fingers tipped with tiny, vestigial claws that grow like upright pegs on the top of bulbous fingertips. Aptly called "Finger-otter" in German, these diminutive otters use their sensitive forepaws constantly to search for prey by touch alone. The Cape or African clawless otter lives south of the Sahara. Its forepaws are not webbed, but have strong clawless fingers which have a truly monkey-like dexterity. Even the thumb shows freedom of movement and can be opposed when picking up or holding down objects. The hindfeet have a small web and the middle toes have claws used for grooming. The fingers are used to probe mud and crevices for crabs and frogs or, in coastal water, octopus. The broad cheek teeth are perfectly adapted to grinding the tough carapace of crustacea, such as rock crabs.

The Giant otter of Brazil, which can measure over 5.9ft (1.8m) overall and weigh up to 66lb (30kg), is among the rarest otters. Its total numbers are unknown, but the population has undoubtedly declined or disappeared over much of its former range. Almost 20,000 skins were exported from Brazil alone in the 1960s. Although the trade declined significantly when bans came into effect in the early 1970s, poaching is still widespread. Prior to overexploitation by pelt hunters, made easy by their quarry's daytime activity and inquisitive habits, the Giant otter was found in most rivers and creeks of the Amazon basin. Its annual life cycle is closely linked to the rise and fall of the water level during the rainy season from April to September. Like other otters, it is an opportunistic feeder, preferring the slower fish that lie camouflaged on the stream or lake bed. These fish move into flooded forests to spawn during the rainy season and the otters follow them until the waters recede again into the creeks a few months later.

The Sea otter of the north Pacific is in many ways unlike other members of the subfamily. Not so slender, and with a relatively short tail, the Sea otter may weigh even more than the Giant otter, and is thus the largest member of the weasel family. It is exclusively marine and rarely comes ashore. Its large, rounded molars are perfectly adapted to crushing sea urchins, abalones and mussels, which it wrestles off rocks with its forepaws. The Sea otter is one of the few tool-using mammals (see p119). Hunted close to extinction for its pelt, the species was protected by an early international agreement in 1911. Total numbers are today probably about 105,000, as a result of the ban and of efforts to reestablish Sea otters in their former range in the 1950s–1970s. Of these, 100,000 or more live from Prince William Sound to the Kurile Islands, and some 2,000 live off California.

Although the Eurasian otter, Smooth-coated otter and Clawless otters have the same distribution in Asia, they are adapted to different diets and habitats, so probably never compete directly with each other. The Eurasian otter prefers quiet streams and lakes away from human disturbance; the Smooth-coated otter can be found in marshes and coastal mangrove swamps; and the *Aonyx* species inhabit shallow estuaries or rice paddies, seldom venturing into deep water.

It seems likely, although the fossil evidence is inadequate to prove it, that *Pteronura* populated South America and *Lutrogale* Asia just before the adaptable and successful *Lutra* otters came along, and that both genera are remnant populations which may well eventually die out.

The social behavior of otters ranges from solitary to family living. *Lutra* species, such as the Eurasian otter, are basically solitary. Although a male and female may pair for several months during the breeding season, there is no strong pair bond between them and the male is dominant during their temporary associations. The Giant, Smooth-coated and Oriental short-clawed otters all live in extended family groups, with strong bonds between breeding pairs—the female is known to be the dominant partner in the first two. The Cape clawless otter is intermediate, in that its members live in pairs; in both *Aonyx* species the male helps raise the young. The Sea otter presents a further social variation: after mating the sexes separate into coastal resting areas that average 108 acres (44ha) in males and 200 acres (80ha) in areas occupied by females with their young; these areas are distinct but may be next to areas of about 74 acres (30ha) patrolled by single males.

Much has been said about the "playfulness" of otters, but this may apply mainly to captive otters with "time on their hands," rather than their wild counterparts. However, wild otters will tunnel through a snow drift or sometimes slide down a mudbank; juveniles may rough-and-tumble ashore or chase each other in water. Play serves to reinforce social bonds, so important to social otters, but also helps young otters perfect their hunting and fighting techniques. Otters "playfully" swimming on a log or rolling in a pile of leaves may in fact be rubbing themselves dry or trying to place a scent mark.

◀ **Manual dexterity** is a striking feature of some otter species. (**1**) The Oriental short-clawed otter (*Aonyx cinerea*) is hand-oriented and will always reach out with its forelimbs for food. (**2**) The Spot-necked otter (*Hydrictis maculicollis*) is mouth-oriented, reaching out with its neck and body for food. (**3**) An Indian smooth-coated otter (*Lutrogale perspicillata*) uses its heavily webbed, but highly dextrous, forepaws to hold a shell to its mouth.

The presence or absence of webbing and claws, as well as manual dexterity, distinguish many otter species. Shown here are forepaws of (**a**) Oriental short-clawed otter, (**b**) Cape clawless otter (*Aonyx capensis*), (**c**) Giant otter (*Pteroneura brasiliensis*), (**d**) Indian smooth-coated otter, (**e**) Spot-necked otter, and (**f**) North American river otter (*Lutra canadensis*).

(**4**) Head of North American river otter. The shape and size of the hairy patch on the nose pad distinguish the different *Lutra* species: (**i**) North American river otter, (**ii**) Eurasian river otter (*L. lutra*), (**iii**) Southern river otter (*L. provocax*), (**iv**) Marine otter (*L. felina*), (**v**) Hairy-nosed otter (*L. sumatrana*), (**vi–viii**) 3 subspecies of the Neotropical river otter: *Lutra longicaudis enudris*, *L. l. platensis* and *L. l. annectens*.

A Tool-using Carnivore

Sea otters are dextrous and particularly versatile in their manipulative skills, even for otters. Most food items are collected by picking them off the bottom or from kelp stalks, but when digging for clams, the otter kicks with its hind flippers to stay close to the bottom while digging rapidly with its forepaws in a circular motion and rooting with its head. It will remain submerged for 30–60 seconds and return to the same hole on three or more successive dives, to enlarge the hole laterally with each dive and retrieve clams as they are encountered.

The Sea otter is the only mammal apart from primates reported to use a tool while foraging. To dislodge abalone they grasp a stone between the mitten-like forepaws and bang it against the edge of the abalone shell. It may require three or more dives to dislodge an abalone; the same stone may be used through 20 or more dives.

Food items are almost always brought to the surface for consumption, from depths of up to 130ft (40m). The Sea otter may then place a stone on its chest and use it as an anvil on which to open mussels, clams, and other shell-encased prey. The stone is carried to the surface in a flap of skin in the armpit and the food item in the forepaws. The stones are usually flat and about 7in (18cm) in diameter. When pounding, the arms are raised to about 90 degrees to the body and the mollusc

Otters are very vocal, with a large repertoire of calls. Different vocalizations readily distinguish *Lutra* from *Lutrogale* and both of these from *Aonyx*. *Lutra* species can be recognized by their unique staccato chuckle (New World) or a twitter (Old World), both given in a context of close proximity between adults or between mother and cubs. The contact call in *Lutra* is a one-syllable chirp, a sound which can carry quite far, whereas in the five other genera the sound is more of a bark with a nasal, guttural quality and the close-contact sound is a humming purr, interspersed with a falling "coo."

Differences in vocalization are to be found mainly in these contact, summons and greeting calls, but similarities, especially in threat and alarm calls, remain. The growl and the inquiring "hah," with minor variations, are common to all the species that have been studied. It is interesting to note that the widely dispersed "giant" otters, that is, *Aonyx* (Asia and Africa), the Sea otter (North Pacific) and Giant otter (South America), share more similarities in vocal repertoire with one another than with species with which they overlap, such as the Eurasian, Indian smooth-coated and Oriental short-clawed otters.

brought down forcefully, so that the hard shell strikes the stone. An uninterrupted series of 2–22 or more blows at about two per second seems enough to crack the shell. It is then bitten and the contents extracted with the lower incisors, which project forward. The otter may roll over in the water between bites to jettison debris and to keep the fur clean. In sandy or muddy areas, where stones are not available, the otter will use one clam or mussel as the tool and bang it against another.

Wild Alaskan Sea otters, unlike captive individuals and members of the Californian population, rarely use anvils, probably because they feed on prey that they can crush in their teeth, such as crab, snails and fish. Sea otters also prey on sea urchins; it is the dye in the urchin's shell that causes the purple color of some Sea otter skeletons.   TRL

Scent marking is a common feature of otter behavior. Only the exclusively marine Sea otter lacks the paired scent glands at the base of the tail which give otters their heavy, musky smell. Scent marking delineates territorial boundaries and communicates information concerning identity, sex, sexual state, receptivity and time elapsed between scenting visits. Otters usually leave single spraints (feces) or urine marks, but the social species also use communal latrines, where the urine and feces are thoroughly mixed and trampled into the substrate by stomping with the hind paws and sometimes (Giant otter) kneading with the forepaws. A pair or a group may clear and scent mark a site together during a bout of feverish activity which leaves an area denuded of vegetation, smelling of scent, feces and urine. The strong, dank odor can be detected near any site which has been visited within the previous several weeks. For a few days the smell is overpowering, but by a week later it is pleasant.

Urine may be dribbled during vegetation marking, when the otter pulls down armfuls of leaves and rubs its body over them. Otters trampling the vegetation cover their fur with the scent they are themselves spreading and later, while resting, they rub themselves against the ground and each other until there is a composite scent characteristic of a pair or even a group.

Recent observations of Brazilian Giant otters show that, when ashore, they use specific scent-marking sites along banks which they clear of all vegetation to a semicircular shape roughly 26ft long and 23ft wide (8m by 7m). On one creek 50 such sites were monitored and at least 23 of them were in areas of perennial vegetation, so that the otters had carefully to keep them clear of grass and fallen twigs. Such sites and their communal latrines are both visual and scent marks, which are further prominently enhanced by trampling the surrounding vegetation and topsoil. One way of marking complements the other—to advertise and to inform the Giant otter that passes by.

That otters are adaptable is evident from their distribution prior to man's emergence. It is tempting to wonder how they will evolve in the next million years—will the Cape clawless otter become more raccoon-like; will the Sea otter resemble a seal more than it does today; will the Brazilian and Smooth-coated otters succumb to the *Lutra* invasion? However, we do not yet even know if otters will survive the 20th century with its pollutants, fur trappers and rampant habitat destruction.   ND

Abbreviations: HBL = head-body length; TL = tail length; wt = weight.
☑ Vulnerable.   ☐ Threatened, status indeterminate.

North American river otter
Lutra canadensis
North American river otter or Canadian otter.

Canada, USA including Alaska. HBL 26–42in; TL 12–18in. Coat: very dark, dusky brown above, almost black to reddish-black or occasionally grayish-brown; lighter, silvery or grayish on belly; throat and cheeks silvery to yellowish-gray. Feet well-webbed, claws strong. Breeding season March–April; gestation 10–12 months with delayed implantation; litter 1–5 (2–3).

European river otter
Lutra lutra
European or Eurasian river otter.

Most of Eurasia S of tundra line, N Africa. HBL 22–27in; TL 14–16in. Coat: brownish-gray to dusky brown (lighter in Asian races); throat buff to cream. Feet well-webbed, claws strong; tail thick at the base. Breeding nonseasonal; gestation 61–65 days; litter 2–5 (2–3).

Marine otter ☑
Lutra felina
Marine otter, chingungo.

Coast and coastal islands of Chile and Peru; exterminated in Argentina. HBL 22–23in; TL 12–14in. Coat: dark brown above, with underside a lighter fawn color. Feet well-webbed, claws strong. Breeding season: no data, may be December–January; gestation 60–70 days; litters average 2.

Southern river otter ☐
Lutra provocax

Argentina, Chile. HBL 22–27in; TL 14–18in. Coat: dark to very dark, burnt umber above; underside a lighter cinnamon color. Claws strongly webbed. Breeding ?

Neotropical river otter
Lutra longicaudis

C and S America from Mexico to Argentina. HBL 20–31in; TL 15–22in. Coat: cinnamon-brown to grayish-brown on back, sometimes with one or more lighter (buff or cream) spots or patches. Claws strong, webbing present. Breeding season varies with locality; gestation ?; litter 1–4 (average 2–3). Taxonomy in the process of revision; here considered to include *L. annectens, L. platensis, L. incarum, L. enudris, L. insularis, L. repanda, L. latidens.*

Hairy-nosed otter
Lutra sumatrana

Sumatra, Java, Borneo, Thailand, Vietnam, Malaysia. HBL 20–32in; TL 14–20in. Coat: very dark brown above, underside very slightly paler; throat sometimes white. Feet well webbed; claws strong. Breeding ?

Spot-necked otter
Hydrictis maculicollis

Africa S of Sahara; absent only from desert areas, such as Namibia. HBL 23–27in; TL 13–18in. Coat: very dark or raw umber above; underside slightly lighter; throat and/or groin usually with irregular patches and spots of cream-white (buff-yellow in juveniles). Webbing to near tips of toes; claws strong. Breeding season variable; gestation ?: litter 1–4.

Indian smooth-coated otter
Lutrogale perspicillata

Discontinuous: Iraq (Tigris river); lower Indus, India, SE Asia, Burma, SW China, Malay Peninsula, Sumatra, Borneo. HBL 26–31in; TL 16–20in. Coat: raw umber to smoky gray-brown throat and cheeks very light gray or almost white. Feet quite large; webbing well-developed and thick; claws strong. Tail tapered, but with slight flattening at sides. Breeding season October or December; gestation 63–65 days; litter 1–4 (2).

Oriental short-clawed otter
Aonyx cinerea
Oriental or Asian short-clawed or small-clawed or clawless otter.

India, Sri Lanka, SE Asia, Indonesia, Borneo, Palawan Islands, S China. HBL 16–25in; TL 10–14in. Coat: burnt umber to dusky brown; throat noticeably lighter, whitish to grayish. Feet narrow, webbed only to about last joint of toes; claws blunt, peg-like, rudimentary. Breeding nonseasonal; gestation 60–64 days; litter 1–6 (2–3).

Cape clawless otter
Aonyx capensis

Africa S of 15° N, from Senegal to Ethiopia S to the Cape; absent only from desert regions of Namibia. HBL 29–37in; TL 16–26in. Coat: dark brown above, sometimes frosted with white or grizzled hair tips; cheeks and neck white. Forefeet virtually unwebbed, looking like pinkish hands; hindfeet similar, but with some webbing; no claws on fingers, inner and outer toes; 3 middle toes have short, peg-like claws. Breeding ?

Giant otter ☑
Pteronura brasiliensis
Giant otter, Brazilian otter.

In all countries of S America except possibly Chile, Argentina and Uruguay. HBL 38–48in; TL 18–25in. Coat: very dark burnt umber, groin never spotted; chin, throat and chest usually marked with cream-colored patches, blotches or spots; muzzle, lips and chin often whitish or spotted white. Claws and webbing very well developed—webs reach to tips of toes and fingers; feet very large, fleshy; tail lance-shaped, widest at middle point. Breeding non-seasonal; gestation 65–70 days; litter 1–5 (2).

Sea otter
Enhydra lutris

Kurile, Aleutian Islands and Gulf of Alaska, California coast; introduced into parts of former range along Pacific coast of N America and USSR. HBL 22–51in; TL 5–13in. Adults dark brown above with head straw-colored; juveniles uniform dark brown; cubs fawn-colored at birth, tail somewhat flattened, short and not markedly tapered; muzzle with thick whiskers. Forefeet small with no obvious toes; hindfeet very large, flipper-like. Breeding season variable, but may be only 4–5 months, gestation about 9 months with delay in implantation; litter 1, rarely 2.

▲ **North American river otters.** Otters range from solitary to highly sociable in their behavior. North American river otters are basically solitary. Like other *Lutra* species, they only pair up for a short while during the breeding season.

◄ **The Giant otter of the Amazon** and other South American waterways has a massive, seal-like head, and a wide flattened wedge of a tail. Its build, unlike that of most otters, does not fit it to terrestrial as well as aquatic life. Giant otters favor shallow creeks, where they prey on characins and small catfish, usually eaten headfirst, firmly clasped between the forepaws, with the elbows resting on the creek bottom.

BADGERS

Subfamilies: Melinae and Mellivorinae
Nine species in 7 genera.
Family: Mustelidae.
Distribution: widespread in Africa, Eurasia, N
America.

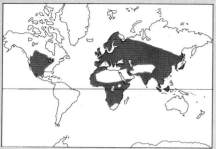

Habitat: chiefly woodlands and forests, also
urban parks, gardens; some species in
mountains, steppe or savanna.

Size: 20in (50cm) and 4.4lb
(2kg) in Ferret badgers to
nearly 39in (1m) and 26lb
(12kg) in the large species.

Gestation: 3½–12 months, including period of
delayed implantation.

Longevity: to 25 years in captivity (not known
in wild).

▼ **Badgers of Southeast Asia.** The tail-less
Malayan stink badger or teledu ABOVE and the
long-tailed Oriental ferret badger BELOW overlap
in Borneo and Java. The ferret badgers are the
smallest species and the only tree climbers.

BADGERS are a widespread group of mostly carnivorous, medium-sized, stocky
mustelids. All species except the Honey badger are classified in the subfamily Melinae
(true badgers), characterized by powerful
jaws and carnassial teeth adapted for
crushing, and molars that are broad, flat and
multicusped. The Honey badger (subfamily
Mellivorinae) is similar in behavior but is
only distantly related; its teeth are not well
developed for crushing, and there are only
four molars in all.

Badgers have powerfully built, wedge-
shaped bodies, with a small head and a short
thick neck. In all but ferret badgers, the tail
is short. The snout is elongate and used for
foraging and, particularly in the Hog badger, is very truncated and mobile. Many
species dig elaborate burrow systems, making use of the relatively long, nonretractile
claws on the forefeet. In the teledu of
Southeast Asia the toes of the forefeet are
even united as far as the roots of the claws.
Badgers walk on their toes (digitigrade) with
a characteristic rolling gait.

Most badgers are nocturnal, although the
Eurasian badger and the American badger
are active in daylight in quiet areas, and the
Honey badger will move by day. In consequence eyesight is relatively poor, the eyes
small and inconspicuous, but the sense of
smell is well developed. The ears are small,

and, relatively, largest in ferret badgers.

Badgers have a tough skin, bearing long
coarse guard hairs, often dull in color but
sometimes with a striking facial pattern.
Like all mustelids, badgers have well-
developed anal glands. The Honey badger's
anal glands secrete a vile-smelling liquid to
deter its enemies, whereas ferret badgers,
the teledu and the Palawan stink badger will
squirt the contents of their anal glands into
the face of an attacker.

Most badgers eat a variety of small vertebrates, invertebrates, fruit and roots. The
American species, the most carnivorous of
all badgers, digs out chipmunks, ground
squirrels, mice and rabbits, it will eat carrion
and invertebrates and also caches food.
Individual American badgers may form
hunting associations with individual
coyotes, but this relationship is poorly understood. In another hunting partnership,
the Honey guide bird's characteristic call
leads the Honey badger to a bees' nest; the
badger breaks open the nest, and both then
feed on the contents.

In many areas the main food of the
Eurasian badger is the earthworm, picked
up from the surface of the ground, where it is
most frequently found during spring and
autumn nights. During wet weather, when
earthworms are available in very large
numbers, the badger will forage on a small

▲ **White facial stripe** and reddish-gray upperparts identify this American badger at the entrance to its burrow. The two northern badgers are sometimes active in daylight hours.

▼ **Familiar black-and-white face markings** of brock the Eurasian badger. Winter inactivity and delayed egg implantation help northern badger populations postpone births until food is more plentiful.

patch of pasture, 100–200sq ft (10–20sq m). slowly quartering the ground (badgers have more difficulty in locating and grabbing earthworms in longer grass, which they consequently avoid). Earthworms are detected when they are immediately below the badger's snout; the badger graps its prey with its incisors—75 percent of earthworms are pulled out entire, the rest have their tails broken off. On favorable nights a Eurasian badger can catch several hundred earthworms in a few hours. In dry weather, when earthworms are less available, badgers follow a more meandering path rather than staying in a few small areas to forage; then badgers also eat fruit, cereals, roots such as pignut, and small mammals.

Male American badgers become sexually mature as yearlings, but 30 percent of females have been found to breed in their first year, when only 4–5 months old. In the

Eurasian badger, both sexes become sexually mature when about one year old, although they may not breed until later. In both species there is a delay between fertilization and implantation of the egg. In northwest Europe, the Eurasian badger may mate between February and October, with most matings in February–May. Implantation of the fertilized egg is delayed 3–9 months, and 1–5 (usually 2 or 3) cubs are born in January–March, depending on latitude. The American badger mates in August or September, with implantation in February and birth of 1–5 cubs (usually 2) in April, or perhaps as late as June at higher altitudes. In both species delayed implantation and their reduced activity in winter may be adaptations to postpone the rearing of the young until more food becomes available. In the Honey badgers of West Asia matings occur in autumn and births after about six months in spring; in Africa matings have been reported in various months. Litter size ranges from 1 to 4.

The Eurasian badger lives in well-defined social groups or clans which usually number up to about a dozen individuals, and occasionally more. During daytime each badger group shares one large main underground burrow (sett) and several smaller outlying setts. Most setts are dug in woodland, in sloping, well-drained soils; some of the larger setts have been in continuous use for many decades or even centuries. Within the territory of each group the outlying setts are randomly distributed; the main setts of adjacent groups are evenly spaced.

Eurasian badger group territories may be as small as 37 acres (15 hectares) or, in moorland, as large as 3,700 acres (1,500 hectares). In large ranges the boundaries are not defended, and the badgers are non-territorial and roaming. On the edges of a smaller, defended territory the often conspicuous pathways are associated with a relatively large number of badger latrines (fewer where one group of badgers has no neighbors). A latrine consists of up to 50 small pits, each up to 4in (10cm) deep. An average-sized latrine might cover an area of 20–55sq ft (2–5sq m). Latrines at territorial boundaries (usually near a conspicuous landmark) are larger than those near setts or elsewhere. Latrine use reaches a peak between February and May, and to a lesser extent in October and November.

Arriving at a latrine, a Eurasian badger may squat, vigorously scratch the ground with its fore- and hindlegs, dig fresh pits with its forelegs, defecate and/or urinate. All these activities are probably accompanied

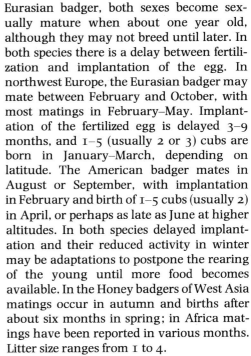

by scent marking from the anal, subcaudal, digital and possibly other glands. Badgers of both sexes will also, particularly at the borders of their territory, squat in a fast action to deposit secretions from the subcaudal gland, the odor of which identifies individuals. Badgers may also mark by rubbing their anal region at a height of 12–16in (30–40cm) up a tree or fencepost, while performing a handstand.

Except for sows with young cubs, badgers usually move around singly. Should a badger meet a member of its own group, one individual will press its anal region against the side or rump of the other, anointing it with secretions from its anal and/or subcaudal glands. But if two strange badgers meet, aggression follows immediately. This may involve the resident chasing and biting the intruder, coupled with growling and hair erection, which may continue for some time after the intruder has been expelled. If two animals from neighboring groups meet on the boundary path, they may simply avoid each other by walking back into their respective ranges.

True badgers (subfamily Melinae) have a long fossil record in the Tertiary (65–2 million years ago) which shows gradually increasing importance of the tubercular teeth at the back of the jaws, the shearing carnassial teeth tending to become reduced—an adaptation to an omnivorous diet. The Eurasian badger (*Meles meles*) may have originated towards the end of this period from the Pliocene genus *Melodon* found in China. The earliest *Meles* fossil in Europe is about 2 million years old (beginning of Pleistocene). The early Middle Pleistocene of Europe was inhabited by badgers similar in appearance to the modern species, and badger remains are common in Middle Pleistocene (including the Honey badger, subfamily Mellivorinae) and late Pleistocene deposits.

The Eurasian badger may be an important reservoir of bovine tuberculosis. In Southeast Asia the teledu is hunted and eaten by some natives, and the American and Eurasian badger are both still hunted for their skins, the coarse hairs being used for shaving and painters' brushes and similar items, and the skins of the Eurasian badger are said to be used in China for rugs.

Although generally shy and retiring animals, both American and Eurasian badgers are occasionally found living in urban areas in close proximity with man. The setts are dug in private gardens or on wasteland, and the diet includes garbage, handouts, fruit and vegetables. In cities badgers emerge later from their setts than elsewhere, but are often seen out in the early morning—on one occasion trotting home along the sidewalk in broad daylight, sixteen feet (5m) behind a couple of joggers! With Eurasian badgers on the rural fringe depleted by snaring and digging, in some places city badgers are replenishing the rural population.   SH

▶ **Searching for its favorite food,** an African ratel or Honey badger attempts to break open a bark-covered behive. Strong claws used for digging, nauseating scent glands and muscular jaws, serve its reputation as a fearless fighter. Its thick skin, hanging like a loose coating of rubber, makes it apparently impervious to tooth, fang or sting and no adversary seems formidable enough for it. Once it has bitten, it never relaxes its grip, snarling and shaking its head, until the victim drops from exhaustion or shakes it loose. When caught young, the Honey badger makes an interesting pet but it may become dangerous when adult, given to sudden bursts of fury, attacking friend or foe indiscriminately.

▼ **The trunk-like, mobile snout** gives the Hog badger of eastern Asia its name.

Abbreviations: HBL = head-body length; TL = tail length; wt = weight.

Honey badger
Mellivora capensis
Honey badger, ratel.

From open, dry savanna to dense forest in Africa, from Cape to Morocco in W and Ethiopia, Sudan and Somaliland in E; Arabia to Russian Turkestan, Nepal and India. Terrestrial, in burrows or among rocks. HBL 24–27in; TL 8–12in; wt up to 26lb. Upper parts from head to tail white (extent of mantle variable, may be absent), sometimes with gray or brown tinge; sides, underparts and limbs pure black. Young rusty brown above.

Eurasian badger
Meles meles
Eurasian or European badger.

Woodland and steppe zones. From Britain, Europe N to S Scandinavia, European Russia up to Arctic Circle, S to Palestine, E to Iran, Tibet and S China. Terrestrial; burrowing. HBL 26–32in; TL 6–8in; wt 26lb (male). 22lb (female). Upper parts gray-black; underparts, legs and feet black. Head and ear tips white, black facial stripe from snout through eyes to behind ears. Male has broader head, thicker neck. Tail narrow, pointed, white or pale in male, broader, grayer in female. Asiatic forms may have brown fur on back.

Hog badger
Arctonyx collaris

Forest zones from Peking in N, throughout S China and Indochina to Thailand, and Sumatra. Terrestrial; burrowing. HBL 22–27in; TL 5–7in; wt 15–31lb. Back yellowish, grayish or blackish; ears and tail white; feet and belly black. Dark facial stripes through eyes, bordered by white stripes merging with white of nape and throat.

Teledu
Mydaus javanensis
Malayan stink badger, teledu.

Mountain zones of Borneo, Sumatra, Java and North Natuna Islands. Terrestrial, occupying simple burrows. HBL 14–20in; TL 2–3in; wt 3–8lb. Coat dark brown or blackish, with white crown to head and either white stripe down back or row of white patches.

Palawan stink badger
Suillotaxus marchei
Palawan or Calamian stink badger.

Palawan and Busuanga (one of Calamian Islands) NE of Borneo. Terrestrial; burrowing. HBL 12–18in; TL 0.5–2in; wt 7lb. Coat dark brown to black above, brown below, with yellowish cap and streak down back, fading at shoulders; muzzle dirty white; anal region hairless and pale-skinned. Heavier-toothed, shorter-tailed and smaller-eared than *Mydaus javanensis*.

American badger
Taxidea taxus

From SW Canada and N central USA, S to Mexico. Terrestrial; burrowing. HBL 16–25in in females, 20–28in in males; TL 4–6in; wt 8–26lb. Females smaller than males. Upper parts grayish to reddish, underparts buff; feet dark brown to black. Central white facial stripe from nose at least to shoulders; black patches on face and cheeks. Chin, throat and mid-ventral region whitish.

Ferret badgers
Three species in the genus *Melogale*.
Indian ferret badger (*M. personata*). India, Nepal, Burma and SE Asia; **Chinese ferret badger** (*M. moschata*), China, Taiwan, Assam, Burma and SE Asia; **Oriental ferret badger** (*M. orientalis*), Java and Borneo. All terrestrial; occasionally climb trees; burrowing. HBL 13–17in; TL 6–9in; wt 4lb. Upper parts pale to dark brown, with white or reddish dorsal stripe; belly paler; face with conspicuous black and white or yellowish pattern.

THE MONGOOSE FAMILY

Family: Viverridae
Sixty-six species in 37 genera.
Distribution: S Italy, France and Iberian Peninsula, throughout Africa, Madagascar, Middle East to India and Sri Lanka, much of C and S China, Hainan, Taiwan, throughout SE Asia to Celebes (civets are the only native carnivores) and the Philippines. Introduced into Pacific and Indian Ocean islands, Kei Islands, W Indies, and Japan.

Habitat: from forests to woodlands, savanna, semidesert and desert.

Size: ranges from the Dwarf mongoose 16in (43cm) long overall weighing on average 11.5oz (320g) to the African civet up to 57in (146cm) long and up to 29lb (13kg) in weight.

True civets, linsangs and **genets** (subfamily Viverrinae): 9 species in 7 genera.
Palm civets (subfamily Paradoxurinae): 8 species in 6 genera.
Banded palm civets (subfamily Hemigalinae): 5 species in 4 genera.
Falanouc (subfamily Euplerinae): *Eupleres goudotii*.
Fossa (subfamily Cryptoproctinae): *Cryptoprocta ferox*.
Fanaloka (subfamily Fossinae): *Fossa fossa*.

Mongooses (subfamily Herpestinae): 27 species in 13 genera.
Madagascar mongooses (subfamily Galidiinae): 4 species in 4 genera.

THE viverrids are one of the most diverse of all carnivore families, but their natural distribution is restricted to the Old World. The family includes all species known as civets, linsangs, genets and mongooses. Several of the eight subfamilies have been at times variously raised to family status; some view the mongooses as a separate family—the Herpestidae—distinct from the other six subfamilies in the Viverridae.

Viverrids so closely resemble the ancestors of carnivores, the Miacoidea, that fossil viverrids are almost indistinguishable from these early Eocene relatives (see p14). The tooth structure and skeletal morphology has barely changed for 40 to 50 million years.

Perhaps the modern viverrids are simply a continuation of this old lineage. However, in spite of their primitive dentition, viverrids have a highly developed inner ear and so present an evolutionary mosaic of primitive and advanced features, making their systematic position uncertain. They are sometimes placed between the weasel and the cat families.

Viverrids vary considerably in form, size, gait (from digitigrade to near-plantigrade) and habits. Most civets and genets resemble spotted, long-nosed cats, with long slender bodies, pointed ears and short legs. However, the binturong (or "Bear cat") resembles a wolverine in build but has long black hair, a very long, thick, prehensile tail and a cat-like head; the African civet is rather dog-like in habits and appearance; the fossa so closely resembles a cat that some scientists once placed it in the cat family as a primitive member; the Otter civets could pass as long-nosed otters; and the falanouc resembles a mongoose with a stretched-out nose and a bushy tail like a tree squirrel. Mongooses vary less in gross form; they have long bodies, short legs and small rounded ears. Most civets have long tails equal to or exceeding their body length. Mongoose tails average half to three-quarters body length.

There is a tendency for males to be slightly larger than females, except for the binturong where females may be 20 percent larger. Females have one to three pairs of teats and males possess a baculum (stiffening bone in the penis). Vision and hearing are excellent.

Viverrids tend to be omnivorous. Most feed on small mammals, birds, reptiles, insects, eggs and fruit. Mongooses take less fruit than civets. Palm civets are almost exclusively fruit eaters.

Mongooses differ from the civet group in a variety of other morphological, behavioral and genetic characters. All mongooses have four or five toes on each foot, nonretractile claws, reduced or absent webbing between toes, and rounded ears placed on the side of the head, rarely protruding above the head's profile; some mongooses have highly developed social systems (see pp144–145); some are active in daylight, others are nocturnal and most are terrestrial. Civets have five toes on each foot, partly or totally retractile claws, webbing between toes, and pointed ears projecting above the head's profile. They tend to be solitary and are primarily nocturnal and tree-dwelling, with a few terrestrial and semiaquatic species. The coat of civets is generally spotted or striped, the ear flaps have pockets (bursae) on the lateral margins and most have a perineal (civet) gland sited near the genitals. Although true mongooses (subfamily Herpestinae) are uniformly colored, lack ear bursae and the perineal gland, Madagascar mongooses (subfamily Galidiinae) have ear bursae and some have perineal glands, and three of the four species are variously striped. In all mongooses the anal glands (containing musk) are well developed.   WCW

▲ **Carefree youth.** Two young Dwarf mongooses play with aloe berries. Dwarf mongooses are the smallest viverrids.

► **Long-nosed, large-eared and spot-coated,** two Common genets wide awake at night in their arboreal home. At the outer edges of the ears can be seen the pockets (bursae) characteristic of civets, linsangs and genets.

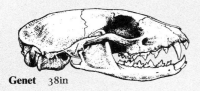

Genet 38in

Mongoose (*Herpestes*) 38in

Binturong 54in

Skulls of Viverrids

Skull forms and dentition vary considerably among viverrids. The facial part of the skull is long, the canine teeth relatively small and the carnassials relatively unprominent. The normal dental formula (see p12) is I3/3, C1/1, P4/4, M2/2 = 40, but the number of molars and premolars may be reduced. Skulls of genets are cat-like (shown here is the Common genet, *Genetta genetta*), while those of civets are similar, but more heavily built. In mongooses, such as *Herpestes* species, the skull is not so long, but more robust. Skulls of Palm civets are also robust, like those of mongooses, with the binturong having a particularly domed form; the primarily vegetarian diet of this species is reflected in its flattened, straight canines, reduced premolars (with one lower pair missing) and molars, and peg-like upper incisors.

CIVETS AND GENETS

Thirty-five species in 20 genera
Family: Viverridae.

Habitat: rain forests to woodlands, brush, savanna and mountains; chiefly arboreal, but also on ground and by riverbanks.

Size: ranges from the African linsang with head-body length 13in (33cm), tail length 15in (38cm) and weight 1.4lb (650g) to the African civet with head-body length 33in (84cm), tail length 17in (42cm) and weight 29lb (13kg); while lighter, the Celebes civet and binturong may be 20 percent longer; some fossas weigh 44lb (20kg).

Coat: various textures; some monochrome species, but dark spots, bands or stripes on lighter ground, and banded tail frequent.

Gestation: 70 days in genets, 80 in African civet, 90 in Palm civets.

Longevity: to about 20 years (in captivity 15 years for Masked palm civet and 34 for genet recorded).

IN the humid night air of the West African rain forest, a loud plaintive cry is repeated like the hooting of an owl; another series of cries penetrates the dark from half a mile away. These are African palm civets, the best known of some 32 species of a group including the civets, linsangs and genets. This diverse assemblage of mostly cat-like carnivores displays a wide range of life-styles and coat markings. Primarily nocturnal foragers and ambush killers, they usually rest in a rock crevice, empty burrow or hollow tree during the day. They are solitary animals, only occasionally forming small maternal family groups.

To man, they are best known for the source of commercial civet oil (see p130). As many are inhabitants of tropical forests, a number of species are finding themselves in an increasingly smaller world—whether this be through the harvesting of lumber in Borneo or the clearing of land for cattle ranching in Madagascar. The IUCN lists as potentially threatened the Otter civet, Jerdon's palm civet and the Large-spotted civet. The status of many of the rarer species is not known. On the other hand the Common palm civet has become so plentiful and accustomed to man that it is frequently found living in and around villages and coffee plantations, and in some areas is considered a pest.

The group is divided into six subfamilies (see BELOW), and here we also discuss the largest (the Viverrinae) in two parts: the true civets and linsangs, and the genets (*Genetta*), since the latter are particularly distinct in their distribution.

The seven southern Asian and one African species of **palm civets** (subfamily Paradoxurinae) are characterized by the possession of semiretractile claws and a perineal scent gland lying within a simple fold of skin. The vocal African palm civet spends most of its time in the forest canopy, where it feeds chiefly on fruits of trees and vines, occasionally on small mammals and birds. Adult males occupy home ranges of over 250 acres (100 hectares) and regularly scent mark trees on the borders of their territory. Dominant males use many kilometers of boughs and vines to make a regular circuit once every 5–10 days through their home ranges. Subordinate males—usually smaller, immature or aged animals—occupy small areas within the range, but avoid dominant males and traverse their ranges at irregular intervals. Eventually, however, when one of these subordinates matures, the dominant male's priority to mate with a female in heat is challenged. Often fatal wounds are inflicted in the ensuing fight and the vanquished landlord, weakened by deep bites, retreats to the ground where he dies or is killed by a leopard or other predator.

One to three females live within the home range of a dominant male and he visits the home range of each for several days as he makes his rounds. For most of the year, females do not tolerate a male staying in the same tree. But in the long rainy season males and females keep track of one another's whereabouts by calling in the darkness. Mating takes place in June over several days, as the pair roost in the same tree. Three months later, 1–3 young are born in a secluded tangle of vines. They are weaned six months later and reach sexual maturity shortly after the second year. Females share their home ranges only with daughters less than two years old. Male offspring emigrate shortly after weaning.

All other species of Palm civets are confined to the forests of Asia. They are skillful climbers, aided by their sharp, curved retractile claws, usually naked soles, and partly fused third and fourth toes which strengthen the grasp of the hindfeet. The Common palm civet is one of the most widespread. Like the Masked palm civet (which, unusually, has no body markings except for the head) it probably forages on the ground for fallen fruit and for animals. The tails of both species are only moderately long and are used to brace the animal during climbing. Most other species have

---

SUBFAMILIES OF CIVETS AND GENETS

Palm civets
Subfamily Paradoxurinae.
Seven species in S Asia, 1 in Africa, including:
African palm civet (*Nandinia binotata*), **Common palm civet** (*Paradoxurus hermaphroditus*), **Masked palm civet** (*Paguma larvata*), **binturong** or Bear cat (*Arctictis binturong*), **Celebes palm civet** [R] (*Macrogalidia musschenbroekii*), and **Small-toothed palm civet** (*Arctogalidia trivirgata*).

Banded palm civets
Subfamily Hemigalinae.
Five species in rain forests of SE Asia.
Banded palm civet [*] (*Hemigalus derbyanus*), **Hose's palm civet** (*Diplogale hosei*), **Owston's banded civet** (*Chrotogale owstonii*), **Otter civet** (*Cynogale bennettii*), and **Lowe's otter civet** [*] (*C. lowei*).

True civets, linsangs and genets
Subfamily Viverrinae.
Nineteen species in Asia, Africa, Arabia, Near

East and SW Europe, including:
African civet (*Civettictis civetta*), **Large Indian civet** (*Viverra zibetha*), **Large-spotted civet** [*] (*V. megaspila*), **Malay civet** (*V. tangalunga*), **Small Indian civet** (*Viverricula indica*), **Banded linsang** [*] (*Prionodon linsang*), **Spotted linsang** [*] (*P. pardicolor*), **African linsang** (*Poiana richardsoni*), **Aquatic genet** or **Congo water civet** (*Osbornictis piscivora*). **Common genet** (*G. genetta*), **Johnston's genet** (*G. johnstoni*), **Forest genet** (*G. maculata*), **Feline genet** (*G. felina*) and **Large-spotted genet** (*G. tigrina*).

Falanouc [V]
Subfamily Euplerinae.
One species in Madagascar, *Eupleres goudotii*.

Fossa [V]
Subfamily Cryptoproctinae.
One species in Madagascar, *Cryptoprocta ferox*.

Fanaloka [V]
Subfamily Fossinae.
One species in Madagascar, *Fossa fossa*.

[*] CITES listed. [R] Rare. [V] Vulnerable.

longer tails. They seem to spend more time foraging in trees. The massively muscular tail of the binturong is prehensile (uniquely among viverrids) and, along with the hind-feet, is used to grasp branches while the forelimbs pull fruiting branches to the mouth. Binturongs have also been reported to swim in rivers and catch fish. The Celebes, Giant or Brown palm civet has very flexible feet with a web of naked skin between the toes. It is an acrobatic climber and lives in steep forested ravines and ridges of central and northeastern Sulawesi. Although not seen for 30 years, a recent report (based on tracks and feces) indicates that it may be fairly common in certain limited areas.

All Palm civets eat a wide variety and a vast quantity of fruit, as well as some rodents, birds, snails and scorpions, which supplement the protein in a diet that is otherwise high in carbohydrates. In Java, the Common palm civet eats the fruits of at least 35 species of trees, palms, shrubs and creepers. Some fruits harmful to humans are eaten without ill effects. The seed of the Arenga palm, for example, has a prickly outer pulp, but it is consumed in large quantities, passing through the digestive tract undamaged. The Small-toothed palm civet has small, flat-crowned premolars which seem to be adaptations for a diet of soft fruit. Nearly all Palm civets are notorious banana thieves and the Common palm civet is also called the Toddy cat for its fondness for the fermented palm sap (toddy) which over much of southern Asia

▲ **Preferring fruit to meat,** the Small-toothed palm civet climbs about the forests of eastern Asia helped by its semiretractile claws, in search of the tree fruits which compose most of its diet.

▶ **Largest of the true civets and linsangs,** the African civet has a tail only half its body length, and does not climb trees but forages on the ground for birds, mammals, reptiles, insects and fruits.

is collected, in bamboo tubes attached to palm trunks, for subsequent human consumption.

There are five species of **banded palm civet** and **otter civet** (subfamily Hemigalinae), all confined to the rain forests of Southeast Asia. The Banded palm civet, the best known, is named for the broad, dark vertical bands on its sides. Very carnivorous, it forages at night on the ground and in trees for lizards, frogs, rats, crabs, snails, earthworms and ants, resting during the day in holes in tree trunks. One to three young are born to a litter and they begin to take solid food at the age of 10 weeks. Hose's palm civet resembles the Banded palm civet in body form, but is distinguished by skull characters and by its uniform blackish-brown color pattern. Owston's banded civet has spots as well as band markings, and seems to be a specialist on invertebrate foods; stomachs of the few specimens now in museums contained only worms.

The Otter civets are the unorthodox members of the group. They have smaller ears, a blunter muzzle, a more compact body and a short tail. There are two species—the Otter civet and Lowe's otter civet—which differ in details of coloration and teeth. Their dense hair, valve-like nostrils and thick whiskers equip Otter civets for life in water, catching fish. The toes are less webbed than in otters, which the Otter civets nevertheless resemble in habit and appearance. Otter civets are fine swimmers and they can also climb trees.

The African civet is the largest and best known of the **true civets and linsangs** (subfamily Viverrinae). It lives in all habitats from dry scrub savanna to tropical rain forest, foraging exclusively on the ground by night and resting by day in thickets or burrows. An opportunistic and omnivorous feeder, it does not disdain carrion, but mammals such as gerbils, spring hares and spiny mice are among its most frequent prey. Ground birds, such as francolins and guinea fowl, are sometimes caught; reptiles, insects and fruits complete the menu. African civets almost always defecate at dung heaps (middens or "civetries") near their route of movement. Civetries are normally less than 5sq ft (0.5sq m) in area and there is evidence that they are located at territorial boundaries serving as contact zones between neighbors. Trees and shrubs which bear fruit eaten by civets are frequently scent marked with the perineal gland, and so to a lesser extent are grass, dry logs and rocks.

Female African civets are sexually mature at one year and may produce as many as two litters a year after a gestation of about

▶ **Representative civets and linsangs.** (1) African linsang (*Poiana richardsoni*) feeding on a nestling. (2) Banded palm civet (*Hemigalus derbyanus*) eating a lizard. (3) Oriental or Malayan civet (*Viverra tangalunga*) with dorsal crest erect. (4) Common palm civet (*Paradoxurus hermaphroditus*) scenting the air. (5) Binturong (*Arctictis binturong*) foraging for fruit, while grasping a branch with its prehensile tail.

▼ **Rare glimpse** of the Celebes civet or Giant civet, which occurs only on the Indonesian island of Sulawesi (Celebes). Few zoo or even museum specimens are known but recent reports indicate that the species is still reasonably common in its very restricted range.

Civet Oil and Scent Marking

The term "civet" derives from the Arabic word *zabād* for the unctuous fluid, and its odor, obtained from the perineal glands of most viverrids, except mongooses. The gland is associated with the genitalia, but it differs in anatomy between species: in Palm civets, it is a simple long fold of skin which produces only a thin film of the scented secretions, while in the three *Viverra* species and in the African civet it is a deep muscular pouch which may accumulate several grams of civet a week. Genet scent has a subtle pleasing odor, but that of the true civets (*Viverra*) is powerful and disagreeable; the scent of the binturong is reminiscent of cooked popcorn. Civetone, which has a pleasant musky odor, is probably the most widespread component, but other compounds, such as scatole, often impart a fetid odor to the secretion.

Scent marking is important in viverrid communication, but the method differs between species. The binturong spreads its scent passively while moving about as the gland touches its limbs and vegetation. Some species scent mark by squatting and then wiping or rubbing the gland along the ground or on a prominent object. The Large Indian civet elevates its tail, turns the pouch inside out and presses it backward against upright saplings or rocks. Like most other species, genets scent mark while squatting, but they also leave scent on elevated objects by assuming a handstand posture.

The close association of the perineal gland and the genitalia suggests that the secretion may have sex-related functions. Indeed, civetone may "exalt" volatile compounds from the reproductive tract of females on heat. Perineal gland scent may also carry information indicative of sex, age and individual identity.

Civet has long played an important role in the perfume industry—it was imported from Africa by King Solomon in the 10th century BC. Once refined, it is cherished within the perfume industry because of its odor, ability to exalt other aromatic compounds and its long-lasting properties. Civet oil also has medicinal uses and has been used to reduce perspiration, as an aphrodisiac and as a cure for some skin disorders. Since the development of synthetic chemical substitutes, the collection of civet oil is not as vital as it once was to the industry. Nevertheless, several East African and Oriental countries still ship large quantities of civet oil each year and, in some instances, civets have been introduced to supply a primitive economic base for poor areas. The animals are kept in small cages, restrained by man-handling, and the scent scraped out with a special spoon.   CW

80 days. One to three young are born, probably in a secluded thicket. The mother nurses them until they are 3–5 months old. They begin to catch and eat insects before weaning and the mother summons them with a chuckling contact call when she wishes to share a rodent or bird that she has caught.

The African civet's Asian relatives probably share many features of its natural history. The similar Large Indian civet, Large-spotted civet, and Malay civet share eye-catching black and white body and back stripes, have a somewhat dog-like body plan and have a crest of long hair overlying the spine which when raised under threat gives the civet an enlarged appearance. Malay civets and Large Indian civets also have the latrine habit of the African civet.

The Small Indian civet has a narrower head and a genet-like build. It lacks a spinal crest, the body spots and tail rings are less well defined than in a genet and the coloration is drab. A skillful predator of small mammals and birds, it stalks its prey like a cat. But it is also an opportunist that feeds on insects, turtle eggs and fallen fruit. It is not a particularly good climber and often lives on cultivated land and near rural villages.

The linsangs are among the rarest, least known and most beautiful members of the group. All three species are small, quick, trimly built, secretive forest animals with darkly-marked torsos and banded tails. They depend almost entirely on small vertebrates for food. The stomachs of four Banded linsangs were found to contain remains of squirrels, Spiny rats, birds, Crested lizards and insects. In all likelihood they live alone and there is good evidence that the African linsang builds leafy nests in trees. The two Asian species, the Spotted linsang and Banded linsang, apparently sleep in nests lined with dried vegetation under tree roots or hollow logs. They lack civet glands and the second upper molar. All three linsangs, like the genets, have fully retractile claws.

The Fishing or Aquatic genet or Congo water civet is a rare and little-known inhabitant of streams and small rivers of the Central African forest block. It feeds on fish and possibly crustaceans, and local people report that it even occasionally eats cassava left to soak in streams before human consumption. It is not particularly specialized for swimming, but probably uses its naked palms to locate fish lurking under rocks and undercut river banks, grabbing them with its retractile claws and killing them with its sharp teeth. cw

Because of their nocturnal habits and cryptic coat patterns, **genets** have been studied mainly not in the wild but in the laboratory. However, recent studies of wild genets in the Serengeti National Park, Tanzania, suggest that their effect on the ecosystem may be considerable.

There are 10 species of these medium-sized, long-bodied and short-legged carnivores, from Africa, Arabia, the Near East and southwest Europe. They all have rows of dark spots along the body, or stripes, which are denser on the upper surfaces, on a light brown or gray background. The tails are ringed and about as long as the body. They have a long face, and pointed muzzle with long whiskers, largish ears, binocular vision, fully retractile claws and five toes on all four feet. In Africa they occupy all habitats except desert, but they prefer areas of dense vegetation. In Spain, Common genets are widespread even in high mountains—one was recently found taking midwinter shelter under the bonnet of a snowplow at the Pyrenean ski resort of Astun, at an altitude of 5,600ft (1,700m).

Common genets may have been imported

► **The perfect killer,** a Common or European genet displays its fine, blade-like set of incisors and long canine teeth. Genets are primarily tree-dwelling carnivores, although small mammals and game birds may be hunted on the ground, and insects and fruit are also taken.

▼ **The Spotted linsang** is the smallest of the true civets and linsangs, and like most of the others is arboreal and has a tail nearly as long as its body.

to Europe as pets by the Moors in the Middle Ages, or they may be a remnant population left after the Gibraltar land bridge was broken. A size variation, with smaller specimens in the north of the range, suggests that the distribution of the Common genet is natural. However, the genets of the Balearic Islands were definitely introduced by man. The subspecies from Ibiza (*G. genetta isabellae*) is smaller than the other European forms; it closely resembles the Feline genet subspecies *G. felina senegalensis* found in Senegal.

Genets are adapted to living in trees, but they often hunt and forage on the ground. Combining speed with stealth, they approach their prey in a series of dashes interrupted by periods of immobility. Their coloration helps them avoid detection, particularly on moonlit nights. Except for Johnston's genet, which may be largely insectivorous, all species are carnivorous; insects and also fruit are a regular addition to their diet. In Africa, small mammals such as rodents make up the major part of the Common genet's diet, while in Spain lizards are also an important item. Feces analysis of Common genets in Spain shows that passerine birds comprise up to half the diet in spring and summer, fruit is important during autumn and winter, rodents (mainly wood mice) are taken all year round, and insects from spring to autumn, though they form a small part of the diet. Some genets stalk frogs at the side of rain pools. One Forest genet has been observed regularly to take bats as they leave their roosts, while in West Africa Forest genets are known to feed on the profuse nectar of the tree *Maranthes polyandra*; the flowers are bat-pollinated, so perhaps the initial attraction was to the bats rather than the nectar.

In Spain, Common genets' activity starts at or just before sunset and ends shortly after dawn. They are inactive for a period during the night, and occasionally there is no morning bout of activity. In daytime the genets only stir to move from one resting site to another. In the Serengeti, the Feline genet, unlike other small carnivores of the plain, is relatively inactive at dawn and dusk and most active around midnight. This pattern of activity may reduce competition for otherwise similar foods.

Genets breed throughout the year, but in many areas there are seasonal peaks, for example April and September for Common genets in Europe. Two to four young are born, after a gestation of about 70 days, in a vegetation-lined nest in a tree or burrow. They are blind at birth and about 5in

(13.5cm) long. Their eyes open after eight days and they venture from the nest soon after that. They are weaned after six months, although they take solid food earlier. Genets are thought to become independent after one year and are sexually mature after two years. Scent marking—by feces, urine and perineal gland secretion—plays an important role in the social life of genets, allowing animals to determine the identity, familiarity, sex and breeding status of other members of their species. Males, and to a lesser extent females, show a seasonal variation in the frequency of marking, with an increase before the breeding season. For most of the time genets live a solitary life.

In Africa, population densities of 2.5–5/sq mi (1–2/sq km) have been reported, with home ranges as small as 0.1sq mi (0.25sq km). Females appear to be more territorial than males, which wander farther. Population density is much lower in Spain. Two male Common genets have been tracked by radio over home ranges of 2sq mi (5sq km), moving quickly over quite long distances, up to 1.9mi (3km) in an hour.

Genets have been widely domesticated and make good pets. In Europe they were kept as rat catchers until they were superseded by the modern domestic cat in the Middle Ages—the cat has a directed killing bite to the nape of the neck which dispatches the prey before it is eaten, whereas genets often hold the prey with all four limbs and eat it alive. In parts of Africa, genets are considered a pest because of their attacks on poultry.   HJH

▲ **Curled up in its hollow-tree home,** a Common genet uses its long tail partly to cover its cat-like head. Before cats became popular, genets were kept as rat-catchers in medieval Europe.

▼ **Half the genet's length is tail.** The Common genet holds its tail straight out behind when it stalks prey at night, usually keeping its body close to the ground. Genets are usually seen singly, sometimes in pairs.

The rare and secretive **falanouc** inhabits wet and low-lying rain forest from east central to northwestern Madagascar. Because of its anatomical peculiarities (see p137) the falanouc is placed in its own subfamily, the Euplerinae, one of three single-species subfamilies of viverrids on Madagascar.

The specialized teeth are used to seize and hold earthworms, slugs, snails and insect larvae. Loath to bite in self-defense, the falanouc employs its sharp claws (long on the forefeet) for this purpose.

The falanouc has a solitary, territorial lifestyle, with a brief consort between mates and a longer mother–young bond that dissolves before the onset of the next breeding season. The base of the tail serves as a fat storage organ for the cold, dry months of June and July when food is in short supply. Subcutaneous fat is deposited in April and May.

The falanouc's reproductive pattern deviates from that of most carnivores: a single offspring is born in summer in a burrow or in dense vegetation. The newborn's well-developed condition suggests that mother and young are highly mobile shortly after birth. An animal born in captivity had open eyes at birth and was able to follow its mother and hide in vegetation when only two days old. It did not take solid food until nine weeks old, but weaned quickly thereafter. These traits make it easy for the young to be constantly close to the mother while she forages for widely dispersed food. Falanouc young develop locomotory and sensory skills very early, but grow and mature at a slightly slower pace than other similar-sized carnivores.

The **fossa** is the largest Madagascan carnivore and large individuals exceed in weight all other members of the civet–mongoose family. Its cat-like head (large frontal eyes, shortened jaws and rounded ears) and general appearance prompted its early classification as a felid. The fossa's resemblance to cats, however, is a result of independent (convergent) evolution. It is quite different from all other viverrids and is the only member of the subfamily Cryptoproctinae. It should not be confused with the fox-like fanaloka (*Fossa fossa*), also from Madagascar.

The fossa evolved on Madagascar to fill the niche of a medium-sized nocturnal, arboreal predator and is an ecological equivalent of the Clouded leopard of Southeast Asia.

The fossa's teeth and claws are adapted to a diet of animals that are captured with the forelimbs and killed with a well-aimed bite. Guinea fowl, lemurs and large mammalian insectivores, such as the Common tenrec, form its basic diet. It is unusual in walking upon the whole foot (plantigrade), not on the toes as in most viverrids.

The fossa lives at low population densities and requires undisturbed forests, which are disappearing fast. Fossas are seasonal breeders, mating in September and October. After a three-month gestation, they give birth to 2–4 young in a tree or ground den. The newborn are quite small—3–3.5oz (80–100g)—compared to other viverrids. Physical development is slow; the eyes do not open for 16–25 days and solid food is not taken for three months. They are weaned by four months and growth is complete at two years, although they do not reach sexual maturity for another two years. Males possess a penis-bone (baculum). Females exhibit genital mimicry of the male, although not as well developed as in the Spotted hyena (see p149).

The **fanaloka** resembles a small spotted fox in build and gait. It is nocturnal and there is also evidence that fanalokas live in pairs, unlike most other viverrids.

The fanaloka mates in August and September and gives birth to one young after a three-month gestation. The young are born in a physically advanced state. The eyes are open at birth and in a few days young are able to follow the mother. The baby is weaned in 10 weeks.

Fanalokas are not particularly good climbers. They rely on hearing and vision to find food and are reported to feed on rodents, frogs, molluscs and sand eels. The fanaloka lives exclusively in dense forests and seems to frequent ravines and valleys. The relationship of the fanaloka to other viverrids is uncertain. Its anatomy differs in many respects from the Banded palm civets with which it was once grouped and it is known only from Madagascar, far away from its presumed south Asian relatives. It is therefore now placed in a separate subfamily, the Fossinae.   cw

◄ **A Large-spotted genet.** Like all genets this southern African species is well adapted to an arboreal way of life. Excellent binocular vision enables it to judge distances accurately as it jumps from branch to branch or pounces on its prey.

▼ **Rarities of Madagascar.** Three unusual viverrids occur on Madagascar. (1) The falanouc (*Eupleres goudotii*) is mongoose-like, has an elongated snout and body, nonretractile claws, and feeds mainly on invertebrates. (2) The fanaloka, or Madagascar civet (*Fossa fossa*), is more fox-like in appearance, has retractile claws and primarily feeds on small mammals, reptiles and amphibia. (3) The fossa (*Cryptoprocta ferox*) has a cat-like head and retractile claws used to capture its prey.

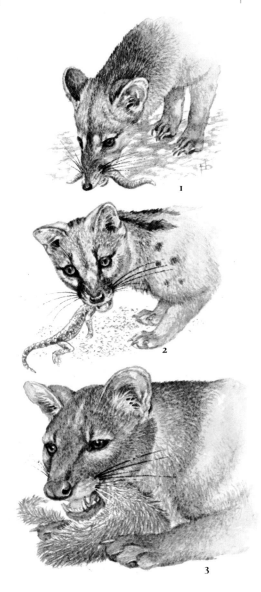

THE 35 SPECIES OF CIVETS AND GENETS

Abbreviations: HBL = head-body length; TL = tail length; wt = weight.

[*] CITES listed. [V] Vulnerable. [R] Rare.

Palm civets
Subfamily Paradoxurinae
(8 species in 6 genera)

Semiarboreal and arboreal; nocturnal. Teeth specialized for a mixed diet or a diet of fruit; carnassial teeth weakly to moderately developed; two relatively flat-crowned molars in the upper and lower jaws of all species. Perineal scent glands present in both sexes of all species, except *Arctogalidia*, in which it is lacking in the male. In *Nandinia* the gland lies in front of the penis and vulva. Claws semiretractile.

African palm civet
Nandinia binotata
African palm civet, Two-spotted palm civet.

From Guinea (including Fernando Póo Island) to S Sudan in the north, to Mozambique, E Zimbabwe and C Angola in south. Arboreal. HBL 20in; TL 22in; wt 7lb. Coat: a uniform olive brown with faint spots; 2 cream spots on the shoulders vary geographically in size and intensity.

Small-toothed palm civet
Arctogalidia trivirgata
Small-toothed or Three-striped palm civet.

Assam, Burma, Thailand, Malayan and Indochinese Peninsulas, China (Yunnan), Sumatra, Java, Borneo, Riau-Lingga Archipelago, Bangka, Bilitung, N Natuna Islands. Arboreal. HBL 20in; TL 23in; wt 5lb. Coat: more or less uniform, varying from silvery to buff to dark brown, sometimes grizzled on head and tail; 3 thin, dark-colored stripes on back, often from base of tail to shoulders; white streak down middle of nose; tail sometimes has vague dark bands; tip sometimes white.

Common palm civet
Paradoxurus hermaphroditus
Common palm civet, Toddy cat.

India, Sri Lanka, Nepal, Assam, Bhutan, Burma, Thailand, S China, Malaya, Indochina, Sumatra, Java, Borneo, Ceram, Kei Islands, Nusa Tenggara (Lesser Sunda Islands) as far E as Timor, Philippines. Semiarboreal. HBL 21in; TL 18in; wt 7lb. Coat: variable from buff to dark brown depending on locality; usually black stripes on back and small to medium spots on sides and base of tail; face mask of spots and forehead streak; spots variable both locally and geographically; tail tip sometimes white.

Golden palm civet
Paradoxurus zeylonensis

Sri Lanka. Arboreal. HBL 20in; TL 18in; wt 7lb. Coat: brown, golden-brown or rusty; spots and stripes barely visible; nap of hair forward on neck and throat; tip of tail sometimes white.

Jerdon's palm civet
Paradoxurus jerdoni

S India, Palni and Nilgiri hills, Tranvancore and Coorg. Arboreal. HBL 23in; TL 20in; wt 8lb. Coat: deep brown, or brown to black with silver or gray speckling; nap of hair as *P. zeylonensis*.

Masked palm civet
Paguma larvata

India, Nepal, Tibet, China (N to Hopei), Shansi, Taiwan, Hainan, Burma, Thailand, Malaya, Sumatra, N Borneo, S Andaman Islands; introduced to Japan. Arboreal. HBL 25in; TL 23in; wt 10lb. Coat: uniform grayish or yellowish-brown to black depending on geographic origin; face dark, but may be marked with a light frontal streak or spots under the eyes and in front of the ears; tip of tail sometimes white.

Celebes civet [R]
Macrogalidia musschenbroekii
Celebes civet, Giant civet, Brown palm civet.

NE and C Sulawesi (Celebes). Semiarboreal. HBL 35in; TL 24in; wt 9lb. Coat: uniform brown with vague darker spots on either side of midline and faint light-colored rings on the tail; hair lighter above and beneath eyes. Cheek teeth of upper jaw arranged in parallel rather than diverging rows.

Binturong
Arctictis binturong
Binturong, Bear cat.

India, Nepal, Bhutan, Burma, Thailand, Malayan and Indochinese Peninsulas, Sumatra, Java, Borneo and Palawan. Arboreal. HBL 30in; TL 29in; wt 17lb (females 20 percent larger). Coat: black with variable amount of white or yellow restricted to hair tips (yellowish or gray binturongs always have black undercoats); hair long and coarse; ears with long black tufts and white margins. Tail heavily built, especially at the base, and prehensile at the tip.

Banded palm civets
Subfamily Hemigalinae
(5 species in 4 genera)

Nocturnal. Second molar large with many cusps. Perineal scent glands present in all species, but not as large as in other subfamilies. Claws semiretractile.

Banded palm civet
Hemigalus derbyanus

Peninsular Burma, Malaya, Sumatra, Borneo, Sipora and S Pagi Islands. Semiarboreal. HBL 21in; TL 12in; wt 5lb. Coat: pale yellow to grayish-buff, with contrasting dark brown markings; face and back with longitudinal stripes; body with about 5 transverse bands extending halfway down flank; tail dark on terminal half, with about 2 dark rings on the base.

Hose's palm civet
Diplogale hosei

Borneo, Sarawak (Mt Dulit to 3,900ft). Semiterrestrial. Dimensions unknown. Coat: uniform dark brown with gray eye and cheek spots; chin, throat and backs of ears white; belly white or dusky gray.

Owston's banded civet
Chrotogale owstoni
Owston's banded or Owston's palm civet.

N of Indochinese Peninsula. Terrestrial. HBL 22in; TL 17in; wt unknown. Coat: similar to Banded palm civet, but with only 4 transverse dark-colored dorsal bands and with black spots on neck, torso and limbs.

Otter civet [*]
Cynogale bennettii
Otter civet, Water civet.

Sumatra, Borneo, Malayan and Indochinese Peninsulas. Semiaquatic. HBL 25in; TL 7in; wt 10lb. Coat: uniform brown, soft dense hair, with faint grizzled appearance; front of throat white or buff-white. First three upper premolars unusually large with high, compressed and pointed crowns; remaining cheek teeth broad and adapted for crushing. Ears small, but designed to keep out water. Feet naked underneath.

Lowe's otter civet
Cynogale lowei

N Vietnam. Semiaquatic. Coat: dark brown above, white to dirty white below from cheek to belly; tail dark brown. Other features as *C. bennettii*.

True Civets, Linsangs and Genets
Subfamily Viverrinae
(19 species in 7 genera)

Teeth usually specialized for an omnivorous diet; shearing teeth well developed; two molars in each side of upper and lower jaws of all except for *Poiana* and *Prionodon*, which have one upper molar. Perineal gland present in all genera except *Prionodon*, presence not certain in *Poiana* and *Osbornictis*. Soles of feet normally hairy between toes and pads.

African linsang
Poiana richardsoni
African linsang or oyan.

Sierra Leone, Ivory Coast, Gabon, Cameroun, N Congo, Fernando Póo Island. Arboreal; nocturnal. HBL 13in; TL 15in; wt 23oz. Coat: torso spotted, stripes on neck; tail white with about 12 dark rings and light-colored tip. Claws retractile.

Spotted linsang [*]
Prionodon pardicolor

Nepal, Assam, Sikkim, N Burma, Indochina. Semiarboreal; nocturnal. HBL 15in; TL 13in; wt 21oz. Coat: light yellow with dark spots on torso and stripes on neck; tail with 8–9 dark bands alternating with thin light bands. Claws retractile.

Banded linsang [*]
Priondon linsang

W Malaysia, Tenasserim, Sumatra, Java, Borneo. Semiarboreal; nocturnal. HBL 16in; TL 13in; wt 25oz. Coat: very light yellow with 5 large transverse dark bands on back; neck stripes broad with small elongate spots and stripes on flank; tail with 7–8 dark bands and black tip. Claws retractile.

Small Indian civet
Viverricula indica
Small Indian civet, rasse.

S China, Burma, W Malaysia, Thailand, Sumatra, Java, Bali, Hainan, Taiwan, Indochina, India, Sri Lanka, Bhutan; introduced to Madagascar, Sokotra and Comoro Islands. Terrestrial; nocturnal/crepuscular. HBL 22in; TL 14in; wt 7lb. Coat: light brown, gray to yellow-gray with small spots arranged in longitudinal stripes on the forequarters and larger spots on the flanks; 6–8 stripes on the back;

neck stripes not contrasting in color as in *Viverra* and *Civettictis*; 7–8 dark bands on tail and tip often light. Claws semiretractile; skin partially bare between toes and foot pads.

Malayan civet
Viverra tangalunga
Malayan or Malay civet, Oriental or Ground civet, tangalunga.

Malaya, Sumatra, Riau–Lingga Archipelago, Borneo, Sulawesi, Karlmata, Bangka, Buru, Ambon and Langkawi Islands, and Philippines. Terrestrial; nocturnal/crepuscular. HBL 26in; TL 12in; wt 8lb. Coat: dark with many close-set small black spots and bars on torso, often forming a brindled pattern; crest, which can be erected, of black hair from shoulder to midtail; black and white neck stripes that pass under throat; white tail bands, interrupted by black crest and black tip. Claws semiretractile.

Large Indian civet
Viverra zibetha

N India, Nepal, Burma, Thailand, Indochina, Malaya, S China. Terrestrial; nocturnal/crepuscular. HBL 32in; TL 17in; wt 19lb. Coat: tawny to gray with black spots, rosettes, bars and stripes on torso, neck with black and white stripes that pass under the throat; erectile spinal crest of black hair from shoulder to rump; tail with complete white bands and black tip. Claws semiretractile.

Large-spotted civet [*]
Viverra megaspila

S Burma, Thailand, formerly the coastal district and W Ghats of S India, Indochina, Malay Peninsula to Penang. Terrestrial; nocturnal/crepuscular. HBL 30in; TL 14in; wt 14lb. Coat: grayish to tawny with small indistinct black or brown spots on the foreparts; large spots on the flanks often fusing into bars and stripes; pronounced black and white neck stripes; spinal crest of erectile black hair from shoulder to rump, bordered on either side by a longitudinal row of spots; tail with 5–7 white bands, most of which do not circle the tail completely, and black tip. Claws not retractile, soles of feet scantily haired between toes and foot pads.

African civet
Civettictis civetta

Senegal E to Somalia in N, through C and E Africa to Zululand, Transvaal, N Botswana and N Namibia in S. Terrestrial; nocturnal/crepuscular. HBL 33in; TL 16in; wt 29lb. Coat: grayish to tawny, torso marked with dark brown or black spots, bars and stripes (degree of striping and distinctness of spots geographically variable); spinal crest from shoulders to tail; tail with indistinct bands and black tip. Claws not retractile; soles of feet bare between toes and foot pads.

Aquatic genet
Osbornictis piscivora
Aquatic genet, Fishing genet, Congo water civet.

Kisangani and Kibale–Ituri districts of Zaire. Semiaquatic; nocturnal. HBL 18in; TL 14in; wt 3lb. Coat: uniform chestnut-brown with dull red belly; chin and throat white; tail uniform dark brown and heavily furred. Claws semiretractile.

Common genet
Genetta genetta
Common genet, Small-spotted genet, European genet.

Africa (N of Sahara), Iberian Peninsula, France, Palestine. Open or wooded country with some cover. HBL 16–20in; TL 14–18in; wt 2–5lb. Coat: grayish-white with blackish spots in rows; tail with 9–10 dark rings and white tip; prominent dark spinal crest.

Feline genet
Genetta felina

Africa S of the Sahara except for rain forest; S Arabian Peninsula. Open or wooded country with some cover. HBL 16–20in; TL 14–18in; wt 2–5lb. Coat: light gray to brownish-yellow with blackish spots in rows; tail with 9–10 black rings and white tip; prominent spinal crest of black hairs; hind legs with gray stripe.

Forest genet
Genetta maculata
(formerly *G. pardina*)

Southern part of W Africa, C Africa, S Africa (except Cape region). Dense forest. Dimensions as *G. genetta*. Coat: grayish to pale brown, more heavily spotted than Common genet; tail black with 3–4 light rings at base and tip dark or light; spinal crest short and can be erected. Relatively long-legged.

Large-spotted genet
Genetta tigrina
Large-spotted genet, Blotched genet, Tigrine genet.

Cape region of S Africa. Woodland and scrub. Dimensions as *G. genetta*. Coat: brown-gray to dirty white with large brown or dark spots; tail relatively long with 8 or 9 black rings and dark tip. Relatively short-legged.

Servaline genet
Genetta servalina
Servaline genet, Small-spotted genet.

C Africa, with restricted range in E Africa. Forest. HBL 16–21in; TL 16–20in; wt 2–4lb. Coat: ocherous and more evenly covered with small blackish spots than other genets; underparts darker; tail with 10–12 rings and white tip. Face relatively long.

Giant genet
Genetta victoriae
Giant or Giant forest genet.

Uganda, N Zaire. Rain forest. HBL 20–24in; TL 18–22in; wt 3–8lb. Coat: yellow to reddish-brown, very heavily and darkly spotted; tail bushy with 6–8 broad rings and black tip; dark spinal crest; legs dark.

Angolan genet
Genetta angolensis
Angolan genet, Mozambique genet, Hinton's genet.

N Angola, Mozambique, S Zaire, NW Zambia, S Tanzania. Forest. Dimensions as *G. genetta*. Coat: largish dark spots in 3 rows each side of erectile spinal crest; tail very bushy with 6–8 broad rings and dark tip; neck striped; hind legs dark with thin gray stripe. Relatively long-haired.

Abyssinian genet
Genetta abyssinica

Ethiopian highlands, Somalia. Mountains. HBL 16–18in; wt 2–3.5lb. Coat: very light sandy-gray with black horizontal stripes and black spotting; tail with 6–7 dark rings and dark tip; back with 4–5 stripes; black spinal crest poorly developed.

Villier's genet
Genetta thierryi
Villier's genet, False genet.

W Africa. Forest and Guinea savanna. Dimensions and coat as for *G. abyssinica* but with chestnut or black spotting which is poorly defined; tail with 7–9 dark rings (first rings rufous) and dark tip.

Johnston's genet
Genetta johnstoni
Johnston's genet, Lehmann's genet.

Liberia. Ground color yellow to grayish-brown; black erectile spinal stripe; rows of large dark spots on sides; tail with 8 dark rings. Skull larger than other genets with overlapping distribution, but greatly reduced teeth suggest largely insectivorous diet.

Subfamily Euplerinae
(1 species)
Falanouc [v]
Eupleres goudotii

East C to NW Madagascar. Rain forest. Terrestrial; solitary. HBL 19–22in; TL 9–10in; wt not known. Coat: light to medium brown; whitish-gray on underside. Snout elongated. First premolars and canines short, curved backward and flattened, for taking small soft-bodied prey. No anal or perineal gland.

Subfamily Cryptoproctinae
(1 species)
Fossa [v]
Cryptoprocta ferox

Madagascar. Rain forest. Arboreal; nocturnal. HBL 27in; TL 25in; wt 21–44lb. Coat: reddish-brown to dark brown. Head cat-like with large frontal eyes, short jaws, rounded ears; tail cylindrical. Carnassial teeth well developed, upper molars reduced (formula I3/3, C1/1, P3/1, M1/1 = 32). Claws retractile. Feet webbed. Anal gland well developed, perineal gland absent. Females show genital mimicry of males.

Subfamily Fossinae
(1 species)
Fanaloka [v]
Fossa fossa
Fanaloka, Madagascar or Malagasy civet.

Madagascar. Dense rain forest. Terrestrial; nocturnal; lives in pairs. HBL 18in; TL 4in; wt 5lb. Coat: brown with darker brown dots in coalescing longitudinal rows; faint dark banding on upperside of short tail. Perineal gland absent.

cw/wcw

MONGOOSES

Thirty-one species in 17 genera
Family: Viverridae.
Distribution: Africa and Madagascar, SW
Europe, Near East, Arabia to India and Sri
Lanka, S China, SE Asia to Borneo and
Philippines; introduced in W Indies, Fiji,
Hawaiian Islands.

Habitat: from forests to open woodland,
savanna, semidesert and desert; chiefly
terrestrial but also semiaquatic and arboreal.

Size: ranges from the Dwarf
mongoose with head-body
length 9.5in (24cm), tail
length 7.5in (19cm) and
weight 11oz (320g) to the
White-tailed mongoose with
head-body length 23in
(58cm), tail length 17in
(44cm) and weight up to
11lb (5kg); some Egyptian
mongooses larger in total
length.

Coat: long, coarse, usually grizzled or brindled;
a few species with bands or stripes.

Gestation: mostly about 60 days, but 42 in the
Small Indian mongoose, 105 in the Narrow-
striped mongoose.

Longevity: to about 10 years (17 recorded in
captivity).

Madagascar mongooses (subfamily Galidiinae)
Four species: **Ring-tailed mongoose** (*Galidia
elegans*), **Broad-striped mongoose** (*Galidictis
fasciata*), **Narrow-striped mongoose**
(*Mungotictis decemlineata*), and **Brown
mongoose** (*Salanoia unicolor*).

African and Asian mongooses (subfamily
Herpestinae)
Twenty-seven species in 13 genera, including:
Dwarf mongoose (*Helogale parvula*), **White-
tailed mongoose** (*Ichneumia albicauda*),
Egyptian mongoose (*Herpestes ichneumon*),
Small Indian mongoose (*H. javanicus*), **Slender
mongoose** (*H. sanguineus*), **Banded mongoose**
(*Mungos mungo*), **Ruddy mongoose** (*H. smithii*),
suricate or **Gray meerkat** (*Suricata suricatta*),
Yellow mongoose or **Red meerkat** (*Cynictis
penicillata*).

▶ **Almost human in pose,** Gray meerkats
(suricates) scan their surroundings in the
Kalahari desert. Like Banded mongooses,
suricates may bunch together to drive off a
potential predator. Group-living mongooses
(unlike most social carnivores) have packs
larger than a single family unit.

AT sunrise a pack of 14 Banded mongooses
leaves its termite mound den. With the
dominant female in the lead, closely followed by the dominant male, they move out
rapidly in single file and then fan out to
search for dung beetles. Contact is maintained by a continuous series of low calls;
from time to time the pitch rises to the
"moving out" call, and the group moves on.
One adult male has remained in the den to
guard the 10 three-week-old young and will
not be seen until they emerge upon the
return of the pack several hours later. Then
the lactating females briefly nurse the young
and several of the younger adults bring
them beetles. Once more the main pack goes
out in search of food, leaving two adults at
the den site to guard the young.

Like other small carnivores, most mongooses are solitary, the only stable social
unit consisting of a mother and her offspring. But some species live in pairs and
several, including the Banded mongoose,
live in groups larger than a single family
unit (see RIGHT and pp144–145). These
group-living mongooses are active during
the day, benefiting perhaps from improved
visual communication, but many solitary
species are nocturnal.

Often the most abundant carnivores in
the locations they inhabit, mongooses are
agile and active terrestrial mammals. The
face and body are long and they have small
rounded ears, short legs and long, tapering
bushy tails.

Most mongooses are brindled or grizzled
and few coats are strongly marked. No
species have spots (unlike civets and genets),
and few have shoulder stripes, and the feet
or legs, and tail or tail tip are often of a
different hue. The Banded mongoose and
the suricate have darker transverse bands
across the back. Among the four Madagascar mongooses, two species have stripes that
run along the body and one has a ringed tail.
Considerable color variation occurs, some-

▼ **Most mongooses are solitary,** like this Slender mongoose and the 10 other *Herpestes* species. Mothers and young form only stable groups. Unlike most mongooses, the Slender mongoose climbs well and will feed on bird eggs and fledglings; birds often mob and dive-bomb this species while ignoring others which pose less of a threat.

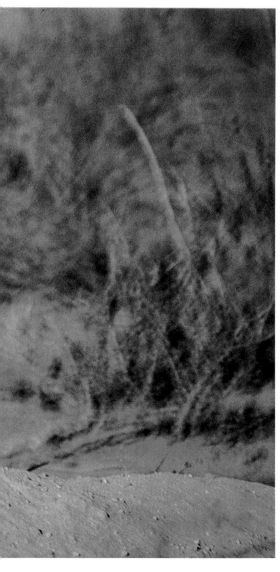

Mutual Defense in Banded Mongooses

Group-living mammals have rarely been seen to rescue one of their number from peril. However, Banded mongooses have several times been observed to come to the aid of their companions, once with dramatic success. On this occasion a Martial eagle swooped down on a foraging pack and seized an adult male, carrying it up to the fork of a tree. The pack then ran to the base of the tree and several mongooses started to climb it with the dominant male in the lead. He reached the branch where the prey struggled in the talons and lunged at the eagle, causing it to loosen its hold. The captured mongoose dropped to the ground unhurt.

Jackals have been observed to catch mongooses from small packs or catch individuals that have become temporarily separated from their pack. But a large pack of Banded mongooses has little to fear from a jackal. When a jackal approaches, the mongooses form a tightly knit bunch and begin to move toward it. Occasionally an individual will stand up to obtain a better view and the bunch gives the appearance of a large animal in constant motion. Jackals retreat from such an apparition and the mongooses may chase and attempt to nip them on the hind legs or tail. (See also p145 for Dwarf mongoose antipredator behavior.)

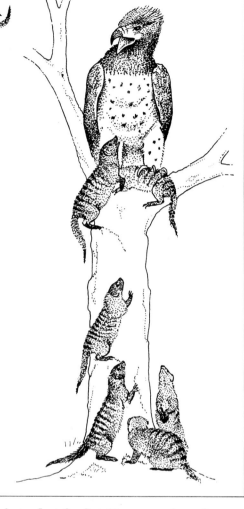

times even within the same species. For example, the Slender mongoose is gray or yellowish-brown throughout most of its range, but in the Kalahari desert it is red, and there is also a melanistic (black) form. Variations usually correlate with soil color, suggesting that camouflage is important to survival.

Most mongooses have a large anal sac containing at least two glandular openings. Scent marking with anal and sometimes cheek glands can communicate the sex, sexual receptivity (estrous condition), and individual and pack identity of the marker.

Ever since Kipling recounted the duel between Riki-tiki-tavi and the cobra, it has been a common assumption in the West that mongooses feed mainly on snakes, but it is unlikely that snakes are sufficiently abundant within the range of any species to predominate in its diet. Most mongooses are opportunistic and feed on small vertebrates, insects and other invertebrates, and occasionally fruits. The structure of the teeth and

feet reflect the diet. Mongooses have from 34 to 40 teeth and those which are efficient killers of small vertebrates, such as *Herpestes* species, have well-developed carnassial teeth used to shear flesh. Their feet have four or five digits each tipped by long, non-retractile claws adapted for digging. The mongoose sniffs along the surface of the ground and when it finds an insect it either snaps it up from the surface or digs it from its underground home.

Some mongooses range over large distances in search of food. On the Serengeti shortgrass plains in Tanzania, packs of Banded mongooses range over approximately 5.8sq mi (15sq km) and may travel over 5.6mi (9km) a day in the dry season. Where food resources are abundant and population density high, ranges and distances of travel are considerably smaller. Banded mongooses in Ruwenzori Park, Uganda, use ranges averaging less than 0.4sq mi (1sq km) and travel about 1.2mi (2km) a day.

In a natural population, some Dwarf mongooses have lived to at least 10 years of age. Wild Small Indian mongooses on Hawaii seldom attain the age of four years, but one Ruddy mongoose lived in captivity for over 17 years.

Most mongooses attain sexual maturity by two years of age. The earliest recorded breeding age is in the Small Indian mongoose, in which females may become pregnant at nine months. Breeding seasons vary depending on environmental conditions. In South Africa the suricate or Gray meerkat and the Yellow mongoose breed only in the warmest (and wettest) months of the year. In western Uganda where the climate is equable and food abundant, Banded mongoose packs usually produce a total of four litters spaced throughout the year, whereas in northern Tanzania, where temperature variation is slight but rainy and dry seasons are pronounced, both Banded and Dwarf mongooses breed only during the months of greatest rainfall when food is most abundant.

In the solitary Slender mongoose, adult males, whose ranges overlap, have a dominance hierarchy. The range of the dominant male includes those of several females and the male moves through these ranges checking scent cues to the females' reproductive condition. There is a brief consortship during the female's estrus. The mother raises the young alone, hiding them from predators. In the group-living Banded and Dwarf mongooses mating occurs within the pack and is regulated by a dominance hierarchy. In most species the young are born sparsely furred and blind, opening their eyes at about two weeks; young of the Narrow-striped mongoose resemble the adults in coloring and have their eyes open at birth.

Mongooses are a widespread and successful group. No species is known to be in danger of extinction, but the most vulnerable are likely to be the four Madagascar mongooses, as a result of destruction of their habitat. The Small Indian mongoose, Yellow mongoose and suricate have been persecuted by man yet are still widespread and abundant (indeed the first named is the most widespread of mongooses). The other two, southern African, species have been shot or gassed in their burrows as rabies carriers. The Small Indian mongoose has also been implicated with rabies and is considered a pest in many parts of the West Indies and Hawaiian Islands because of its attacks on chickens and native fauna. It was first introduced into the West Indies in the 1870s and to the Hawaiian Islands in the 1880s, in an attempt to control rats in the sugarcane plantations. Although it is sometimes said that the Small Indian mongoose is responsible for causing the extinction of many native West Indian birds and reptiles, there is no proof of this. On many islands this mongoose is still an important predator on harmful rodents and its economic status should be considered separately on each island.

The Egyptian mongoose was considered sacred by the ancient Egyptians and mongoose figures have been found on the walls of tombs and temples dating back to 2800 BC. Interestingly, the well-known Welsh myth of Llewellyn and his dog (faithful pet saves child from predator, runs all bloodied to welcome his master home and is killed on presumption that he has killed the child) apparently passed through many cultures from an early Indian tale of the Brahmin, the snake and the mongoose.   JR

▶ **Representative mongoose species.** (1) White-tailed mongoose (*Ichneumia albicauda*), largest of the mongooses. (2) Bushy-tailed mongoose (*Bdeogale crassicauda*), Kenyan subspecies, sniffing the air in typical mongoose "high-sit" posture. (3) Ring-tailed mongoose (*Galidia elegans*) in fast, active trot. (4) Dwarf mongoose (*Helogale parvula*) adult feeding beetle to juvenile. (5) Selous' mongoose (*Paracynictis selousi*) in low, sitting posture. (6) Narrow-striped mongoose (*Mungotictis decemlineata*). (7) Egyptian mongoose (*Herpestes ichneumon*) preparing to break open an egg by throwing it between its legs onto a rock. (8) Marsh mongoose (*Atilax paludinosus*) scent-marking a stone by the "anal drag" method.

▼ **In typical "tripod" posture,** a pair of Banded mongooses check for predators, before setting off on a foraging trip. While defense and care of young are social activities in some species, no mongooses hunt cooperatively, although they may forage together.

THE 31 SPECIES OF MONGOOSES

African and Asian Mongooses
Subfamily Herpestinae

Do not have ear bursa (pocket in flap of each ear) or perineal scent gland. Females have 2 or 3 pairs of teats.

Marsh mongoose
Atilax paludinosus
Marsh or Water mongoose.

Gambia east to Ethiopia, south to S Africa. Nocturnal; solitary; semiaquatic. HBL 19in; TL 13in; wt 8lb. Coat: dark brown to black. No webbing between toes. Bare heel pad present.

Bushy-tailed mongoose
Bdeogale crassicauda

Mozambique, Malawi, Zambia, Tanzania, Kenya. HBL 17in; TL 9in; wt 4lb. Nocturnal; solitary. Coat: dark brown to black; Kenya subspecies lighter in color. Four (not 5) toes on each foot.

Black-legged mongoose
Bdeogale nigripes

Nigeria to N Angola, C Kenya, SE Uganda. Nocturnal; solitary. HBL 24in; TL 15in; wt 7lb. Coat: light gray to brown with yellow to white-tipped guard hairs; fur dense and

▼ **Foraging as a pack,** a group of Banded mongooses search for insects on the Serengeti shortgrass plains. Packs contain up to 40 individuals.

short; belly and tail white; chest and legs black. Four (not 5) toes on each foot.

Alexander's mongoose
Crossarchus alexandri
Alexander's or Congo mongoose.

Zaire, W Uganda, Mt Elgon, Kenya. Diurnal; group-living. HBL 16in; TL 10in; wt 3lb. Coat: mainly brown with black feet. Nose elongate and mobile.

Angolan mongoose
Crossarchus ansorgei

N Angola, SE Zaire. Diurnal; group-living. HBL 12in; TL 8in; wt about 2lb. Coat: brownish with black legs and tail tip. Nose not elongate.

Kusimanse
Crossarchus obscurus
Dark mongoose, kusimanse, or Long-nosed mongoose.

Sierra Leone to Cameroun. Diurnal; group-living. HBL 13in; TL 8in; wt 2lb. Coat: dark brown to black. Nose elongate and mobile.

Yellow mongoose
Cynictis penicillata
Yellow mongoose or Red meerkat.

S Africa, Namibia, S Angola, Botswana. Diurnal; lives in pairs or family groups. HBL 12in; TL 8in; wt 1.7lb. Coat: tan-yellow to orange, speckled with gray; chin and tail tip white. Ears large.

Pousargues' mongoose
Dologale dybowskii
Pousargues' or Dybowski's or African tropical savanna mongoose.

NE Zaire, Central African Republic, S Sudan, W Uganda. Diurnal. HBL 11in; TL 8in; wt 0.8lb. Coat: dark brown grizzled with tan; feet and legs black.

Dwarf mongoose
Helogale parvula

Ethiopia to northern S Africa, west to N Namibia, Angola and Cameroun. Diurnal; group-living. HBL 9in; TL 7in; wt 0.7lb. Coat: varies from grayish-tan to dark brown with fine grizzling. Bare heel pad present. (Includes *H. hirtula*.)

Short-tailed mongoose
Herpestes brachyurus

Malaysia, Sumatra, Java, Philippines. Nocturnal/crepuscular; solitary. HBL 19in; TL 8in; wt 3lb. Coat: brown to black with guard hairs banded with brown to red; legs black. (Includes *H. hosei*.)

Indian gray mongoose
Herpestes edwardsi

E and C Arabia to Nepal, India and Sri Lanka. Diurnal; solitary. HBL 17in; TL 15in; wt 3.3lb. Coat: gray to light brown, finely speckled with black.

Indian brown mongoose
Herpestes fuscus

S India and Sri Lanka. Solitary. HBL 15in; TL 12in; wt 3.5lb. Coat: blackish-brown to sandy-gray, speckled with black. Feet darker than head and body.

Egyptian mongoose
Herpestes ichneumon
Egyptian mongoose or ichneumon.

Most of Africa except for the Sahara and C and W African forest regions and SW Africa; Israel, S Spain and Portugal. Mainly diurnal; solitary. HBL 22in; TL 20in; wt 8lb. Coat: grizzled gray, with black tail tuft.

Small Indian mongoose
Herpestes javanicus
Small Indian mongoose or Javan gold-spotted mongoose.

N Arabia to S China and Malay Peninsula; Indochina; Sumatra; Java; introduced into the W Indies, Hawaiian Islands, Fiji. Diurnal; solitary. HBL 15in; TL 10in; wt 1.7lb. Coat: either light brownish-gray, speckled with black to dark brown (in arid regions), or red speckled with black and gray (wet tropical regions). (Includes *H. auropunctatus*.)

Long-nosed mongoose
Herpestes naso

SE Nigeria to Gabon and Zaire. Nocturnal: solitary. HBL 22in; TL 16in; wt 7lb. Coat: dark blackish-brown grizzled with buff; crest of hairs from nape to shoulders. Nose distinctly elongate.

Cape gray mongoose
Herpestes pulverulentus

S Angola, Namibia, S Africa. Crepuscular/nocturnal; solitary. HBL 13in; TL 13in; wt 1.6lb. Coat: brownish-gray to gray, speckled with black (black and red forms known); tail with dark brown to black tip; feet dark brown. Bare heel pad.

Slender mongoose
Herpestes sanguineus

Africa S of Sahara. Diurnal; solitary. HBL 13in; TL 12in; wt 1.3lb. Coat: varies from gray, yellowish or reddish-brown to red, tail tip black. Bare heel pad.

Ruddy mongoose
Herpestes smithi

India, Sri Lanka. Nocturnal; solitary. HBL 18in; TL 16in; wt 4lb. Coat: light brownish-gray to black, speckled with white and red; feet dark brown; tail tip black.

Crab-eating mongoose
Herpestes urva

S China, Nepal, Assam, Burma, Indochinese Peninsula, Taiwan, Hainan, Sumatra, Borneo, Philippines. Nocturnal; solitary. HBL 20in; TL 12in; wt 7.4lb. Coat: dark brown to gray; legs black; white strip from mouth to shoulder; tip of tail light-colored.

Stripe-necked mongoose
Herpestes vitticollis

S India, Sri Lanka. Diurnal and crepuscular; solitary. HBL 21in; TL 14in; wt 6.3lb. Coat: grizzled-gray, tipped with chestnut; legs and feet dark brown or black; tail tip black; black stripe from ear to shoulder. Bare heel pad.

White-tailed mongoose
Ichneumia albicauda

Subsaharan Africa except for the C and W African forest regions and SW Africa; S Arabia. Nocturnal; solitary. HBL 23in; TL 17in; wt 9lb. Coat: gray, grizzled with black and white; tail usually white but may be dark in some areas; legs and feet black.

Liberian mongoose
Liberiictis kuhni

Liberia. Diurnal; group-living. HBL 16in; TL 8in; wt 5lb. Coat: blackish-brown with dark brown stripe bordered by two light brown stripes from ear to shoulders.

Gambian mongoose
Mungos gambianus

Gambia to Nigeria. Diurnal; group-living. HBL 12in; TL 7in; wt 4lb. Coat: grizzled-gray and black; black stripe on side of neck contrasting with buffy white throat.

Banded mongoose
Mungos mungo

Africa S of Sahara excluding Congo and SW Africa. Diurnal; group-living. HBL 13in; TL 9in; wt 4lb. Coat: brownish-gray; feet dark brown to black; tail with black tip; dark brown bands across back.

Selous' mongoose
Paracynictis selousi
Selous' mongoose or Gray meerkat.

S Angola, N Namibia, N Botswana, S Zambia, Zimbabwe, northern S Africa, Mozambique. Nocturnal; solitary. HBL 18in; TL 14in; wt 3.7lb. Coat: brown, speckled with gray and black; tail white-tipped. Four toes on each foot.

Meller's mongoose
Rhynchogale melleri

S Zaire, Tanzania, Malawi, Zambia, C and N Mozambique. Nocturnal; solitary. HBL 18in; TL 15in; wt 6lb. Coat: reddish-brown; grizzled with tan; legs dark brown.

Suricate
Suricata suricatta
Suricate, meerkat, Gray meerkat, stokstertje.

Angola, Namibia, S Africa, S Botswana. Diurnal; group-living. HBL 11in; TL 7in; wt 2lb. Coat: tan to gray with broken dark brown bands across back and sides; head and throat grayish-white; eye rings and ears black; tail tip black.

Madagascar Mongooses
Subfamily Galidiinae

Ear bursa present; perineal scent gland present in *Galidia* and *Galidictis*. Females have one pair of teats.

Ring-tailed mongoose
Galidia elegans

Madagascar. Diurnal; live in pairs. HBL 14in; TL 11in; wt 2lb. Coat: light tan to dark red-brown; 5-7 dark bands on tail. Bare heel pad.

Broad-striped mongoose
Galidictis fasciata
Broad-striped mongoose or Madagascar banded mongoose.

Madagascar. Nocturnal; live in pairs. HBL 14in; TL 11in; wt ? Coat: grayish-brown with broad dark longitudinal stripes from nape to slightly beyond base of tail.

Narrow-striped mongoose
Mungotictis decemlineata

W Madagascar. Diurnal; live in pairs. HBL 13in; TL 11in; wt 1.7lb. Coat: brownish-gray with speckling on back and sides; 10-12 narrow reddish-brown to dark brown longitudinal stripes on back and sides. Bare heel pad.

Brown mongoose
Salanoia unicolor

E Madagascar. Diurnal; live in pairs. HBL 9in; TL 6in; wt 1.7lb. Coat: reddish-brown, speckled with black and tan.

JR/WCW

Pack Life of the Dwarf Mongoose

Anti-predator behavior and care of young in a small carnivore

The young female Dwarf mongoose dashed back and forth between the two adult males, grooming them and being groomed in turn. In human terms, she seemed ecstatic.

"Bonnie" had been observed from birth. Seven weeks after the death of her mother, Bonnie's pack had been joined by two older females from an adjacent one. Bonnie—then 19 months old—had left (or perhaps been evicted) the same day. For the following month she lived alone, gradually moving out of her natal range. Now she had just found a pack whose dominant breeding female had died in the previous month.

A few minutes later the rest of the pack returned. Bonnie was chased by a juvenile female (the only female in the pack), and again that night she slept alone. But four days later she was a fully integrated member and she was subsequently observed mating with the alpha (dominant) male. Now, as the new alpha female, her reproductive potential was high.

Dwarf mongooses, the smallest viverrids, are also one of the few social species in the family. Each Dwarf mongoose pack contains one dominant breeding pair and this status, once achieved, is often maintained for many years. One alpha female who lost a hindleg to a predator, maintained her position, in spite of this handicap, and bred until she disappeared from the pack three years later.

In addition to the dominant breeding pair (usually the oldest male and female in the group) a Dwarf mongoose pack typically contains young animals born in the pack, and immigrants. Packs of over 20 occur, but the average pack size at the start of the birth season is about eight. Because mongooses are basically monogamous it is usually essential for a mongoose to attain alpha status in order to raise offspring successfully. Some mongooses remain in their natal packs and eventually accede to alpha status when older individuals of the same sex die. The wait may be a long one.

Females are more likely than males to remain in the pack in which they were born. But most mongooses eventually emigrate. They may meet other emigrants of the opposite sex to form a new pack on unoccupied territory or they may join an already existing pack. If they are lucky, as in the case of Bonnie, they can find a pack with no older mongooses of their sex and are thus able to breed at an early age. Many join packs with one or two older residents of their sex and frequently have to contend with some aggression from these. Emigrants move to packs with fewer older mongooses of the same sex, thereby increasing their repro-

ductive potential. Male takeovers, involving the forced expulsion of resident males, have been recorded. In one instance six males joined an adjacent pack, driving out the two resident adult males and mating with the females. The following year all but the dominant male emigrated again (in groups of three and two) and took over two further packs.

In most group-living mammals, females remain in their natal groups while males emigrate. Dwarf mongooses are unusual in that *both* sexes commonly transfer between groups. Intergroup transfer helps prevent inbreeding and also can result in reinforcement of the pack. Members of small vulnerable packs which have lost breeding members can quickly improve their chances of survival through accepting immigrants.

◄ ▼ **Anti-predator behavior** is an important function of group life in the Dwarf mongoose. From vantage points, such as termite mounds, the vigilant mongoose either checks the sky for aerial predators LEFT or scans the surrounding terrain BELOW (1).

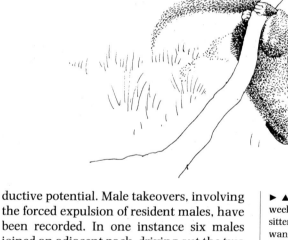

► ▲ **Care of young is shared.** In the early weeks of life the young are guarded by baby-sitters (2) who will rapidly retrieve any that wander away from the den TOP RIGHT. Mutual grooming BOTTOM RIGHT is frequent between members of a pack, particularly between individuals of the opposite sex, and wrestling and chasing play (3) are common among young animals. (4) Scent marking with the anal glands: from such scent marks other mongooses can identify the individual, its sex and pack membership.

for survival from numerous aerial and ground predators. Dwarf mongooses spend a large proportion of their time scanning from vantage points such as termite mounds—the alpha male is particularly active in this role. On detection of a predator, a loud series of alarm calls warns all pack members, who scatter to shelter.

Increased efficiency in care of young is also an important benefit of group living. Young Dwarf mongooses are cared for co-operatively. For the first few weeks, when confined to a breeding den, they are usually guarded by one or more babysitters of either sex while the rest of the pack forages. Changeovers occur frequently throughout the day, allowing all individuals some time to feed. Helpers collect beetles and other insects and carry them to the young, and they groom and play with the young. If any wander away, the babysitters soon retrieve them. Potential predators, such as Slender mongooses, are chased from the den site. The mother spends less time at the den than other pack members, allowing her maximum foraging time.

According to the theory on kin selection, it should (all else being equal) be close relatives that give the greatest amount of aid to the young mongooses. Yet in Dwarf mongooses unrelated immigrants are frequently good helpers and may contribute more than older siblings. Why do they do this? They may receive long-term pay-offs, such as benefiting eventually from the anti-predator responses of young they help to rear. Also, immigrants are likely eventually to breed in the packs they have joined, and the young they have helped to raise may later aid them by babysitting and feeding the immigrants' own young. JR

It appears that the group-living mongooses have followed a different evolutionary route to sociality from the large social carnivores such as lions, hyenas, African wild dogs and wolves. These species hunt cooperatively, which has probably been the most important selective pressure promoting their group life. In contrast, the group-living mongooses all feed primarily on invertebrates and find their food individually. For them predation has probably been the chief selective pressure favoring group living.

Antipredator behavior (as in the Banded mongoose, see p139) is an important benefit of group life. A group of animals is more likely to spot a predator than a single individual. The small Dwarf mongoose must rely on early warning and fleeing to cover

THE HYENA FAMILY

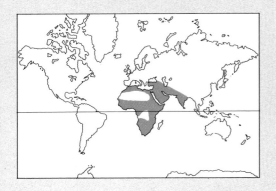

Family: Hyaenidae
Four species in 3 genera.
Distribution: Africa except Sahara and Congo basin, Turkey and Middle East to Arabia, SW USSR and India.

Habitat: chiefly dry, open grasslands and brush.

Size: the Spotted hyena can attain 176lb (80kg) and is one of the largest carnivores. The smallest member of the family is the aardwolf (see below).

Spotted hyena *Crocuta crocuta*
Brown hyena *Hyaena brunnea*
Striped hyena *Hyaena hyaena*
Ardwolf *Proteles cristatus*

Aardwolf   53in

T HE true hyenas (subfamily Hyaeninae) have thickset muzzles with large ears and eyes, powerful jaws and big cheek teeth to deal with a carnivorous diet. They walk on four-toed feet with five asymmetrical pads and nonretractile claws. The tail is long and bushy (less so in the Spotted hyena). The aardwolf is mainly insectivorous. It has retained five toes on its front feet, and its unusual dentition has led some authorities to place it in a separate family (Protelidae). Anatomy, chromosomes and blood proteins, however, clearly indicate a close relationship to hyenas.

Despite the resemblance to canids, on the basis of comparative anatomy the nearest relatives are the Viverridae (civets etc, see pp126–127), and they probably evolved from a civet-like creature similar to *Ictitherium* before the early Miocene era (26 million years ago). The earliest fossil hyenids are from Europe and Asia; one, with advanced bone-crushing teeth, dates from the late Miocene era (10 million years ago).

Abbreviations: HBL = head-body length; TL = tail length; HT = height; wt = weight.
⬚ CITES listed. ⬚ Vulnerable.

Spotted hyena

Crocuta crocuta
Striped or Laughing hyena.

Africa S of Sahara, except southern S Africa (exterminated) and Congo basin. Prefers grassland and flat open terrain. HBL 47–55in; HT 27–35in; TL 10–12in; wt 110–176lb.
Coat: short, dirty yellow to reddish, with irregular dark brown oval spots; short, reversed, erectile mane; tail with brush of long black hairs. Gestation: 110 days. Longevity: up to 25 years (to 40 in captivity).

Brown hyena ⬚

Hyaena brunnea
Brown hyena or Beach wolf or Strand wolf.

S Africa, particularly in W; now absent from extreme S. Prefers drier, often rocky areas with desert or thick brush. HBL 43–51in; HT 25–33in; TL 8–10in; wt 77–110lb.
Coat: dark brown; white collar behind ears, well-developed dorsal crest extending to mantle of long (10in) hairs on back and sides; underside lighter; dark horizontal stripes on legs; face hair short, near-black. Gestation: about 84 days. Longevity: up to 13 years in captivity.

Striped hyena ⬚

Hyaena hyaena

E and N Africa (not Sahara or West south of 10°N) through Mid East to Arabia, India, SW USSR. Habitat similar to Brown hyena. HBL 39–47in; HT 25–31in; TL 10–13in; wt 66–88lb. Coat: medium to long (5–10in), gray to yellowish gray with numerous black stripes on body and legs; woolly winter underfur present in colder areas; muzzle, throat underside, neck and two cheek stripes black; mane usually black-tipped; tail long, of uniform color. Gestation: about 84 days. Longevity: up to 24 years in captivity.

Aardwolf

Proteles cristatus.

Southern Africa north to S Angola and S Zambia; E Africa from C Tanzania to NE Sudan. Open country and grassland; also savanna, scrub and rocky areas. Head to tail-tip 33–41in; HT 15–20in; wt 17–26lb. Coat: buff, yellowish-white or rufous; erectile mane and black dorsal stripe to black tail tip; three vertical black stripes on body, 1–2 diagonal stripes across fore- and hindquarters; irregular horizontal stripes on legs, darker towards feet; throat and underparts paler to gray-white; sometimes black spots or stripes on neck; woolly underfur with longer, coarser guard hairs. Gestation: 59–61 days. Longevity: to 13 years in captivity.

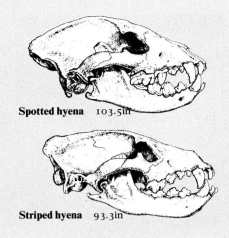

Spotted hyena 103.5in

Striped hyena 93.3in

Skulls of Hyenids

The skulls of hyenids are robust and long. All members of the family have a complete dental formula (see p12) of I3/3, C1/1, P4/3. M1/1 = 34 (among carnivores only members of the cat family have fewer teeth). However, in the insectivorous aardwolf the cheek teeth are reduced to small, peg-like structures spaced widely apart and often lost in adults, leaving as few as 24 teeth.

The more massive skulls of hyenas have relatively short jaws which give a powerful grip. Hyena skulls suggest two distinct trends of adaptation. That of the Spotted hyena is highly specialized for crushing large bones and cutting through thick hides, which other large carnivores are unable to consume and digest. The bone-crushing premolars are relatively large and the carnassial teeth are used almost solely for slicing or shearing.

In Brown and Striped hyenas the corresponding premolars are smaller and the carnassials do the job of crushing and chopping as well as shearing. These differences relate to the smaller species' dependence on a greater range of food items, including insects, wild fruit and eggs as well as carrion and prey.

◄ **A massive head,** jaws with large bone-crushing teeth, and powerful forequarters are hallmarks of the Spotted hyena, largest member of the hyena family. Here a pack consumes a carcass.

▼ **Feeding behavior of the four species of the hyena family.** (1) Aardwolf upwind of termites, listening for sound of termites eating. (2) Striped hyena scavenging from a carcass. (3) Spotted hyena pack cooperatively hunting down a zebra. (4) Brown hyena juveniles playing at the den while an adult approaches with its kill, a Bat-eared fox.

The radiation of hyenas was linked to an increase in open habitats where their dog-like characteristics would have evolved. Probably the hyenas and the North American "hyena-dogs" evolved their special dentition as an adaptation to the availability of the tougher portions of their kills left uneaten by the great saber-toothed cats. As the saber-tooths declined during the early Pleistocene (2 million years ago) so did the hyena-dogs and hyenas, including the massive Cave hyenas (*Crocuta crocuta spelaea*) almost twice the size of those which exist today.

The evolutionary lines leading to *Crocuta* and *Hyaena* appear, from the fossil record, to have been separate since the Miocene era. The ancestor of the aardwolf may have diverged even earlier.

Hyenas are master scavengers, able to consume and digest items that would otherwise remain untouched by mammals. Their digestive system is fully equal to their unusual tastes; the organic matter of bone is digested completely and indigestible items (horns, hooves, bone pieces, ligaments and hair) are regurgitated in pellets, often matted together with grass. This specialized means of eliminating waste is probably the reason why hyenas do not regurgitate food

for their young like many other carnivores.

The ability of Spotted hyenas to hunt is as impressive as their scavenging—a single hyena can catch an adult wildebeest weighing up to 380lb (170kg) after chasing it for 3mi (5km) at speeds of up to 37mph (60km/h). Other individuals may join the chase and in the Ngorongoro Crater in East Africa more than 50 members of the same clan (group) may eventually feed together. Zebra hunting, on the other hand, involves parties of 10–15 hyenas. Following a hunt, Spotted hyenas feed voraciously—a group of 38 hyenas has been seen to dismember a

zebra in 15 minutes, leaving few scraps. When competition is less intense, feeding is more leisurely

The three species of hyena possess the well-developed forequarters and sloping backline, anal pouch and dorsal mane that are common to all hyenids. The biggest, the Spotted hyena, exceeds in size all other carnivores except the four largest bears and three largest felids. There is no overlap in the distribution of the Brown hyena and the Striped hyena, but members of both species come into contact with Spotted hyenas.

Male and female hyenas look alike, but

How Hyenas Communicate

Hyenas are often called "solitary," a label which obscures the fact that their social systems are among the most complex known for mammals. Spotted hyenas employ elaborate meeting ceremonies and efficient long-range communication by scent and sound. Brown and Striped hyenas lack loud calls but their meeting rituals and uses of scent are no less complex. (In the aardwolf, only the scent-marking approaches that of the other members of the family in complexity: see p151.)

Scent marking. One of the most distinctive features of all hyenids is the anal pouch. This remarkable organ lies between the rectum and the base of the tail and can be turned inside out. It is particularly large in the Brown hyena, which secretes two distinct pastes from different glands lining the pouch. As the animal moves forward over a grass stalk (**1**) with its pouch extruded, a white secretion is deposited first, followed by a black one a few centimeters above it (**2**). Chemical analysis of the pastes reveals consistent differences between individuals, while pastes deposited at different times by one animal are extremely similar. Scent marks are placed throughout the territory (averaging 1.4 marks per mile) but the rate of pasting nearly doubles in the vicinity of borders. Striped and Spotted hyenas deposit a single creamy paste, usually on a grass stalk at about hyena nose height.

In Spotted hyenas, which have the least developed anal pouches, aggression is evident during bouts of communal scent marking at border latrines, where pasting is accompanied by defecation and vigorous pawing of the ground with the front feet, which carry glands between the toes.

Submission and aggression signals. All three species turn the anal pouch inside out during encounters with other animals. In Brown and Striped hyenas, this occurs when meeting other members of the species, whereas in the Spotted hyenas, anal gland protrusion is strongly linked with signs of aggression, for example, when approaching lions or rival hyenas.

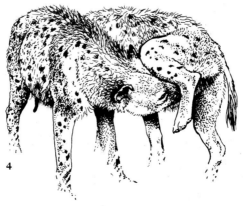

The link between anal glands and aggression in Spotted hyenas is of special interest because, unlike other hyenas, the anal region is not presented for inspection during meeting ceremonies. It seems that the selective advantage of reducing tension while re-establishing social bonds between partners after separation has resulted in two types of display. Meeting ceremonies in Brown and Striped hyenas involve varying degrees of erection of the dorsal crest (**3**), sniffing of the head and body, protrusion and inspection of the anal pouch and rather lengthy bouts of ritual fighting in which areas of the neck or throat of a subordinate are bitten and held or shaken. In Spotted hyenas, on the other hand, greeting includes mutual sniffing and licking of the genital area and erect penis or clitoris as the animals stand head to tail with one hind leg lifted (**4**). This display is very different from the state of the sex organs at mating; it is conspicuous in cubs, and the Spotted hyena which initiates contact is almost invariably lower in the dominance hierarchy. Clearly the function is one of appeasement.

Whoop calls. Even when moving alone, Spotted hyenas maintain some direct contact with their fellows. They respond to sounds which are only audible to humans with the aid of an amplifier and headphones. Calls audible to the unaided human ear include whoops, fast whoops, yells and a kind of demented cackle that gives this species its alternative name of Laughing hyena. Whoop calls, in particular, are well suited to long-range communication as they carry over several kilometers; each call is repeated a number of times, which helps the listener to locate the caller, and each hyena has a distinctive voice. Woodland Spotted hyenas of the Timbavati Game Reserve in the Transvaal, South Africa, will frequently either answer tape-recordings of their companions or casually approach the loudspeaker. If, however, recordings of strange hyenas are played, the residents will often arrive in groups at the run, calling excitedly, with their manes raised, tails curled high and anal glands protruding. Infant hyenas will answer the pre-recorded whoops of their mothers, but not those of other clan hyenas. SKB

▲ **Caching food in mud.** Spotted hyenas frequently bury excess food in muddy pools; they have a good memory for such caches, to which they will return when hungry.

◀ **Brown hyenas** TOP are mainly solitary scavengers and they search for suitable food mostly at night. This one is removing the last morsels of flesh from a buffalo skull.

▲ **A frustrated Spotted hyena** attempts to break open an ostrich egg by biting it. He has already tried stamping and rolling on it. He may fail to open it, although hyenas are often known to eat eggs.

the female Spotted hyena, which is socially dominant, is heavier than the male (4–12 percent). Her sexual organs mimic those of the male so exactly that it is often difficult to be certain of an animal's sex. Male Brown and Striped hyenas are 7–12 percent heavier than females, and the sexual organs are conventional.

Spotted hyenas will eat almost anything, but in the wild 90 percent or more of their food comes from mammals heavier than 44lb (20kg), which they mostly kill for themselves. The frequency of hunting depends on the availability of carrion; Spotted hyenas will loot the kills of other carnivores, including lions. Group feeding is often noisy, but rarely involves serious fighting. Instead, each hyena gorges up to 33lb (15kg) of flesh extremely rapidly. Pieces of a carcass may be carried away to be consumed at leisure or, occasionally, stored underwater.

Brown and Striped hyenas are mainly scavengers, but a significant proportion of their diet consists of insects, small vertebrates, eggs and also fruits and vegetables, an important source of water. Lone animals follow a zigzag course with the head lowered, but frequently turn and sniff into the wind. Small prey are chased and grabbed, but hunting is often unsuccessful. Apart from scents, sounds of other predators and their dying prey also attract hyenas, who may either wait patiently to scavenge,

or drive off the true owner. When Brown or Striped hyenas discover a large source of food they usually first remove portions to the safety of temporary caches among bushes, in long grass or down holes. One Brown hyena watched in the Kalahari removed all 26 eggs from an ostrich nest in one night and returned later to feed on the hidden eggs.

Striped and Brown hyenas spend more time searching for food that is widely scattered and in small clumps than do Spotted hyenas, and their social system is adapted accordingly. At one extreme, extended family groups of 4–14 Brown hyenas in the Kalahari may share territories of 90–200sq mi (230–540sq km) and be active for over 10 of the 24 hours each day, during which time they may travel an average 20mi (30km) or more. By contrast, clans of Spotted hyenas in the Ngorongoro Crater may number 30–80, occupy territories of 26–105sq mi (10–40sq km), and be active for just 4 hours of the day, traveling only some 6mi (10km). Marked variation occurs within the species: the less numerous Spotted hyenas of the Transvaal Lowveld woodlands usually move alone (60 percent of sightings) or in pairs (27 percent).

Hyenas show interesting differences in the care of young. In Spotted hyenas this is the sole responsibility of the mother, but clan females usually raise their offspring in a communal den where narrow interconnecting tunnels allow the infants to escape predators, which may include males of their own species. Up to three infants, usually twins, are born; they are relatively well developed at birth, with a coat of uniform brown. At first the young are called to the surface by the mother to suckle. Movements away from the den develop very slowly and suckling may continue for up to 18 months, but at no stage is food carried back to the den for the benefit of the offspring. Quite the reverse is true in both *Hyaena* species. Female Brown hyenas will suckle infants which are not their own, while in both species adults and subadults of both sexes carry food to related offspring. Rather surprisingly, in Brown hyenas this helping usually excludes the father, as mating seems to be only by nomadic males who wander through separate territories of each extended family. The mating system of Striped hyenas is unknown, but both species generally produce 2–4 blind and helpless young, similar in color to the adults. They are suckled for up to 12 months. Female Striped hyenas have six teats. Brown and Spotted hyenas four.

It seems that the success of Spotted hyenas is ensured through individual and cooperative hunting and sharing of food between adults. Cooperation also extends to communal marking and defense of the territory, in which both sexes play a similar role, whether or not they are related. Competition within the clan can, however, be intense. The system of communication shows adaptations which reduce aggression and coordinate group activities (see p148). Such competition probably provided the selection pressure whereby females evolved their large size and dominant position, which in turn relates also to levels of testosterone in the blood that are indistinguishable from those of the male. Thus female Spotted hyenas are able to feed a small number of offspring alone and protect them from the more serious consequences of interference by other hyenas, particularly unrelated males. Although Brown and Striped hyenas are known to share a large carcass, group members rarely eat together, so direct competition for food is avoided. Indirect competition is offset by the fact that the residents of a territory are nearly always related, and this may explain also why they cooperate in raising young. Through communal rearing of a larger number of infants and the efficient use of small food items, Brown and Striped hyenas are better able to exploit their harsher environments than is the Spotted species. On the other hand they are less well equipped to deal with large prey and their numbers may be kept down by direct competition with Spotted hyenas where their ranges overlap.   SKB

The **aardwolf** is a delicate, shy, nocturnal animal seldom seen in the wild. Its highly specialized diet consists primarily of a few species of Snouted harvester termites (*Trinervitermes* spp). The aardwolf appears to locate its prey mainly by sound, but the strong-smelling defense secretions of the soldier termites probably provide an additional stimulus. The termites are licked up by rapid movements of the long tongue. Because of the sticky saliva that covers the tongue, large amounts of soil may be ingested with the food.

The behavior of aardwolves—including their time of greatest activity, foraging method and social system—is influenced by their dependence on termites. For most of the year aardwolves' periods of activity are similar to those of the Snouted harvesters, which are poorly pigmented and cannot tolerate direct sunlight, so emerge during the late afternoon and at night. The termites

forage in dense columns and an aardwolf can lick up a great number at a time. Certain seasonal events, such as the onset of the rains in East Africa and the cold temperatures of midwinter in southern Africa, appear to limit the termites' own foraging activity. Then aardwolves often find an alternative food in the larger harvester termite *Hodotermes mossambicus*, which is heavily pigmented and may be found by day in large, locally distributed foraging parties. These termites are not the preferred food source throughout the year because they are mainly active in winter and foraging individuals are spaced much further apart than in *Trinervitermes*. Insects other than termites or ants, and very occasionally small mammals, nestling birds and carrion may be eaten but constitute a very minor part of the diet.

Aardwolves are solitary foragers. This is because *Trinervitermes* forage in small dense patches 1–3ft (25–100cm) across scattered over a wide area. One adult pair of aardwolves usually occupies an area of 0.4–0.8sq mi (1–2sq km) with their most recent offspring. An intruding aardwolf may be chased away up to 1,300ft (400m), and serious fights take place if the intruder is caught. Most fights take place during the mating season, when they occur once or twice a week. Fights are accompanied by hoarse barks or a type of roar with the mane and tail hairs fully erected.

When food is short, the territorial system may be relaxed, allowing several individuals from up to three different territories to forage simultaneously in the same area (usually on *Hodotermes*) without any serious conflict. However, even within a family, interaction

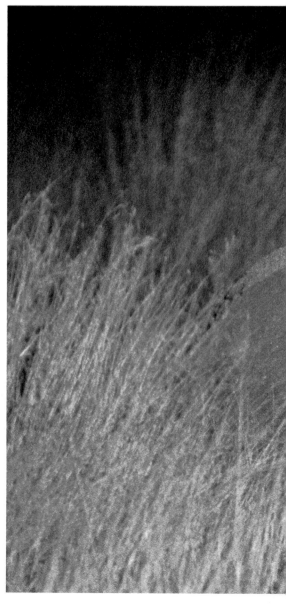

between individuals is abrupt, and there are no elaborate greeting procedures such as exist among hyenas.

Apart from aggressive encounters, the territorial system appears to be maintained also by a system of marking. Both sexes possess well-developed anal glands which can be extruded to leave a small black smear about 0.2in (5mm) long on grass stalks, usually close to a termite mound. Aardwolves mark throughout the night as they move across their territories feeding. When deep within their territories, they mark about once every 20 minutes only, pasting over old marks or around dens and middens, where they may mark up to five times during one visit. The frequency of marking goes up dramatically when they are feeding or simply patrolling the territory boundary, with marking occurring about once every 160 feet (50m). In this way an individual may deposit 120 marks in two hours. This high frequency of marking is most pronounced during the mating season.

An aardwolf family group may have over 10 dens and as many middens scattered throughout its territory. Defecations at middens are usually preceded by the aardwolf digging a small hole and concluded by it scratching sand back over the feces.

The dens may be old aardvark or porcupine dens, or crevices in rocks but often they are holes of a typical size, which the aardwolves may have dug themselves (aardwolf is Afrikaans for "earth wolf") or have enlarged from springhare holes. Aardwolves often visit old dens, but use only one or two at a time and change dens after a month to six weeks. During cold weather they usually go down the den and sleep a few hours after sundown. In summer they rest outside the den entrance at night and go underground during the day.

Although the aardwolf has a strict territorial system, many males are inclined to wander through adjacent territories, particularly during breeding, when resident as well as neighboring males may mate with females. The cubs, usually 2–4, are born in spring or summer. They are born with their eyes open but are helpless and spend about 6–8 weeks in the den before emerging. During the first few months while the cubs are still in the den, the male may spend up to six hours a night looking after the den while the female is away foraging. At about three months the cubs start foraging for termites, accompanied by at least one parent; by the time they are four months old they may spend much of the night foraging alone. They generally sleep in the same den as their mother, while the male may sleep there or in another den. At the start of the next breeding season the cubs often wander far beyond their parents' territory and by the time the next generation of cubs start foraging away from the dens, most of the subadults have emigrated from the area. Despite this annual movement of aardwolves, recolonization of suitable areas is severely limited by man's persecution. Aardwolves have been shot in the mistaken belief that they prey upon livestock, while in some areas they may also be killed for their meat, which is considered a delicacy, or for their pelts. An aardwolf may consume up to 200,000 termites in one night, and since Harvester termites can be serious pests on livestock farms (particularly during drought), the species deserves protection in those areas where it is threatened.

PRKR/SKB

◄ **Pack power.** A large pack of Spotted hyenas is quite capable of intimidating much larger carnivores. Here at least ten hyenas are driving away three lionesses from a kill.

▼ **Licking the platter clean,** an aardwolf feeds direct from a mound of the snouted harvester termite *Trinervitermes trinervoides*, the chief food of aardwolves in South Africa. The aardwolf's long, mobile tongue is covered with sticky saliva and large papillae to help in licking up the termites.

BIBLIOGRAPHY

The following list of titles indicates key reference works used in the preparation of this volume and those recommended for further reading. The list is divided into two categories, those on mammals in general and those specifically devoted to the carnivores.

General

Boyle, C. L. (ed) (1981) *The RSPCA Book of British Mammals*, Collins, London.

Corbet, G. B. and Hill, J. E. (1980) *A World List of Mammalian Species*, British Museum and Cornell University Press, London and Ithaca, N.Y.

Dorst, J. and Dandelot, P. (1972) *Larger Mammals of Africa*, Collins, London.

Grzimek, B. (ed) (1972) *Grzimek's Animal Life Encyclopedia*, vols 10, 11 and 12. Van Nostrand Reinhold, New York.

Hall, E. R. and Kelson, K. R. (1959) *The Mammals of North America*, Ronald Press, New York.

Harrison Matthews, L. (1969) *The Life of Mammals*, vols 1 and 2. Weidenfeld & Nicolson, London.

Honacki, J. H., Kinman, K. E. and Koeppl, J. W. (eds) (1982) *Mammal Species of the World*, Allen Press and Association of Systematics Collections, Lawrence, Kansas.

Kingdon, J. (1971–82) *East African Mammals*, vols I–III, Academic Press, New York.

Morris, D. (1965) *The Mammals*, Hodder & Stoughton, London.

Nowak, R. M. and Paradiso, J. L. (eds) (1983) *Walker's Mammals of the World* (4th edn), 2 vols, Johns Hopkins University Press, Baltimore and London.

Vaughan, T. L. (1972) *Mammalogy*, W. B. Saunders, London and Philadelphia.

Young, J. Z. (1975) *The Life of Mammals: their Anatomy and Physiology*, Oxford University Press, Oxford.

Carnivores

Bekoff, M. (1978) *Coyotes: Biology, Behavior and Management*, Academic Press, New York.

Bertram, B. C. (1978) *Pride of Lions*, Charles Scribner, New York.

Dominis, J. and Edey, M. (1968) *The Cats of Africa*, Time-Life, New York.

Eaton, R. L. (1974) *The Cheetah: the Biology, Ecology, and Behavior of an Endangered Species*, Van Nostrand Reinhold, New York.

Ewer, R. F. (1973) *The Carnivores*, Weidenfeld & Nicolson, London.

Fox, M. W. (ed) (1975) *The Wild Canids: their Systematics, Behavioral Ecology, Evolution*, Van Nostrand Reinhold, London and New York.

Frame, G. & L. (1981) *Swift and Enduring: Cheetahs and Wild Dogs of the Serengeti*, Dutton, New York.

Guggisberg, C. A. W. (1961) *Simba: the Life of the Lion*, Howard Timmins, Cape Town.

Herrero, S. (ed) (1972) *Bears: their Biology and Management*, IUCN Publ. New Series no. 23, Morges, Switzerland.

Hinton, H. E. and Dunn, A. M. S. (1967) *Mongooses: their Natural History and Behaviour*, Oliver & Boyd, Edinburgh and London.

Kruuk, H. (1972) *The Spotted Hyena: a Study of Predation and Social Behavior*, University of Chicago Press, Chicago.

Lawick, H. van and J. van Lawick-Goodall (1970) *The Innocent Killers*, Collins, London.

Mech, L. D. (1970) *The Wolf: the Ecology and Behavior of an Endangered Species*, Natural History Press, Garden City, New York.

Mountfort, G. (1981) *Saving the Tiger*, Michael Joseph, London.

Neal, E. G. (1977) *Badgers*, Blandford, Poole, Dorset.

Pelton, M. R., Lentfer, J. W. and Stokes, G. E. (eds) (1976) *Bears: their Biology and Management*, IUCN Publ. New Series no. 40, Morges, Switzerland.

Powell, R. A. (1900) *The Fisher: Life History, Ecology and Behavior*, University of Minnesota Press, Minneapolis.

Schaller, G. B. (1967) *The Deer and the Tiger*, Chicago University Press, Chicago.

Schaller, G. B. (1972) *The Serengeti Lion: a Study of Predator-Prey Relations*, University of Chicago Press, Chicago.

Verts, B. J. (1967) *The Biology of the Striped Skunk*, University of Illinois Press, Urbana.

Wrogemann, N. (1975) *Cheetah Under the Sun*, McGraw-Hill, Johannesburg.

GLOSSARY

Adaption features of an animal which adjust it to its environment. Adaptations may be genetic, produced by evolution and hence not alterable within the animal's lifetime, or they may be phenotypic, produced by adjustment on the part of the individual and may be reversible within its lifetime. NATURAL SELECTION favors the survival of individuals whose adaptations adjust them better to their surroundings than other individuals with less successful adaptations.

Adaptive radiation the pattern in which different species develop from a common ancestor (as distinct from CONVERGENT EVOLUTION whereby species from different origins became similar in response to the same SELECTIVE PRESSURES).

Adult a fully developed and mature individual, capable of breeding, but not necessarily doing so until social and/or ecological conditions allow.

Aerobic deriving energy from processes that require free atmospheric oxygen, as distinct from ANAEROBIC processes.

Agouti a grizzled coloration resulting from alternate light and dark barring of each hair.

Air sac a side-pouch of the larynx (the upper part of the windpipe), used in some primates and male walruses as resonating chambers in producing calls.

Alloparent an animal behaving parentally toward infants that are not its own offspring: the shorthand jargon "HELPER" is most commonly applied to alloparents without any offspring of their own and it can be misleading if it is used to describe any non-breeding adults associated with infants, but which may or may not be "helping" by promoting their survival.

Alveolus a microscopic sac within the lungs providing the surface for gaseous exchange during respiration.

Amphibious able to live on both land and in water.

Anaerobic deriving energy from processes that do not require free oxygen, as distinct from AEROBIC processes.

Anal gland or sac a gland opening by a short duct just inside the anus or on either side of it.

Arteriole A small artery (ie muscular blood vessel carrying blood from the heart), eventually subdividing into minute capillaries.

Arterio-venous anastomosis (AVA) a connection between the ARTERIOLES carrying blood from the heart and the VENULES carrying it back to the heart.

Axilla the angle between a forelimb and the body (in humans, the armpit)

Baculum (*os penis*, or penis bone) an elongate bone present in the penis of certain mammals.

Binocular form of vision typical of mammals in which the same object is viewed simultaneously by both eyes; the coordination of the two images in the brain permits precise perception of distance.

Biomass a measure of the abundance of a life-form in terms of its mass, either absolute or per unit area (the population densities of two species may be identical in terms of the number of individuals of each, but due to their different sizes their biomasses may be quite different).

Biotic community a naturally occurring group of plants and animals in the same environment.

Blastocyst see IMPLANTATION.

Boreal region a zone geographically situated south of the Arctic and north of latitude 50°N: dominated by coniferous forest.

Brindled having inconspicuous dark streaks or flecks on a gray or tawny background.

Cache a hidden store of food: also (verb) to hide for future use.

Canopy a fairly continuous layer in forests produced by the intermingling of branches of trees; may be fully continuous (closed) or broken by gaps (open). The crowns of some trees project above the canopy layer and are known as emergents.

Carnassial (teeth) opposing pair of teeth especially adapted to shear with a cutting (scissor-like) edge: in extant mammals the arrangement is unique to Carnivora and the teeth involved are the fourth upper premolar and first lower molar.

Carnivore any meat-eating organism; alternatively, a member of the order Carnivora, many of whose members are carnivores.

Caudal gland an enlarged skin gland associated with the root of the tail. Subcaudal: placed below the root; supracaudal: above the root.

Cecum a blind sac in the digestive tract of a mammal, at the junction between the small and large intestines, particularly well developed in some specialized leaf-eaters.

Cerebral cortex the surface layer of cells (gray matter) covering the main part of the brain, consisting of the cerebral hemispheres.

Cerrado (central Brazil) a dry savanna region punctuated by patches of sparsely wooded vegetation.

Chaco (Bolivia and Paraguay) a lowland plains area containing soils carried down from the Andes: characterized by dry deciduous forest and scrub, transitional between rain forest and pampas grasslands.

Chromatin materials in the chromosomes of living cells containing the genes and proteins.

Class taxonomic category subordinate to a phylum and superior to an order (see TAXONOMY).

Clavicle the collar-bone

Coniferous forest forest comprising largely evergreen conifers (firs, pines, spruces etc), typically in climates either too dry or too cold to support deciduous forest. Most frequent in northern latitudes or in mountain ranges.

Convergent evolution The independent acquisition of similar characteristics in evolution, as opposed to possession of similarities by virtue of descent from a common ancestor.

Crepuscular active in twilight.

Cryptic (coloration or locomotion) protecting through concealment.

Cusp a prominence on a cheek-tooth (premolar or molar)

Deciduous forest temperate and tropical forest with moderate rainfall and marked seasons. Typically, trees shed leaves during either cold or dry periods.

Delayed implantation see IMPLANTATION.

Den a shelter, natural or constructed, used for sleeping, for giving birth and raising young, and/or in winter; also the act of retiring to a den to give birth and raise young, or for winter shelter.

Dental formula a convention for summarizing the dental arrangement whereby the numbers of each type of tooth in each half of the upper and lower jaw are given: the numbers are always presented in the order: incisor (I), canine (C), premolar (P), molar (M). The final figure is the total number of teeth to be found in the skull. A typical example for Carnivora would be I3/3, C1/1, P4/4, M3/3 = 44.

Dentition the arrangement of teeth characteristic of a particular species.

Desert areas of low rainfall, typically with sparce scrub or grassland vegetation or lacking vegetation altogether.

Digit a finger or a toe.

Digital glands glands that occur between or on the toes.

Digitigrade method of walking on the toes without the heel touching the ground (cf PLANTIGRADE).

Dimorphism the existence of two distinct forms (polymorphism = several distinct forms); the term "sexual dimorphism" is applied to cases where the male and female of a species differ consistently in, for example, shape, size, coloration and armament.

Disjunct or **discontinuous distribution** geographical distribution of a species that is marked by gaps. Commonly brought about by fragmentation of suitable habitat, especially as a result of human intervention.

Dispersal the movements of animals, often as they reach maturity, away from their previous home range (equivalent to EMIGRATION). Distinct **dispersion**, that is, the pattern in which things (perhaps animals, food supplies, nest sites) are distributed or scattered.

Display any relatively conspicuous pattern of behavior that conveys specific information to others, usually to members of the same species; can involve visual and/or vocal elements, as in threat, courtship or "greeting" displays.

Distal far from the point of attachment of origin (eg tip of tail).

Diurnal active in daytime.

Dominant see HIERARCHY.

Dormancy a period of inactivity; many bears, for example, are dormant for a period in winter: this is not true HIBERNATION, as pulse rate and body temperature do not drop markedly.

Ecology the study of plants and animals in relation to their natural environmental setting. Each species may be said to occupy a distinctive ecological NICHE.

Ecosystem a unit of the environment within which living and nonliving elements interact.

Ecotype a genetic variety within a single species, adapted for local ecological conditions.

Elongate relatively long (eg of canine teeth longer than those of an ancestor, a related animal, or than adjacent teeth).

Emigration departure of animal(s), usually at or about the time of reaching adulthood, from the group or place of birth.

Enzootic concerning disease regularly found within an animal population (endemic applies specifically to people) as distinct from EPIZOOTIC.

Epizootic a disease outbreak in an animal population at a specific time (but not persistently, as in ENZOOTIC); if an epizootic wave of infection eventually stabilizes in an area, it becomes enzootic.

Erectile capable of being raised to an erect position (erectile mane).

Esophagus the gullet connecting the mouth with the stomach.

Estrus the period in the estrous cycle of female mammals at which they are often attractive to males and receptive to mating. The period coincides with the maturation of eggs and ovulation (the release of mature eggs from the ovaries). Animals in estrus are often said to be "on heat" or "in heat." In primates, if the egg is not fertilized the subsequent degeneration of uterine walls (endometrium) leads to menstrual bleeding. In some species ovulation is triggered by copulation and this is called **induced ovulation**, as distinct from spontaneous ovulation.

Family a taxonomic division subordinate to an order and superior to a genus (see TAXONOMY).

Fast ice sea ice which forms in polar regions along the coast, and remains fast, being attached to the shore, to an ice wall, an ice front, or over shoals, generally in the position where it originally formed.

Feces excrement from the bowels; colloquially known as droppings or scats.

Feral living in the wild (of domesticated animals, eg cat, dog).

Fermentation the decomposition of organic substances by microorganisms. In some mammals, parts of the digestive tract (eg the cecum) may be inhabited by bacteria that break down cellulose and release nutrients.

Fissipedia (Suborder) name given by some taxonomists to modern terrestrial carnivores to distinguish them from the suborder Pinnipedia which describes the marine carnivores. Here we treat both as full orders: the Carnivora and the Pinnipedia.

Fitness a measure of the ability of an animal (with one genotype or genetic make-up) to leave viable offspring in comparison to other individuals (with different genotypes). The process of NATURAL SELECTION, often called survival of the fittest, dermines which characteristics have the greatest fitness, that is, are most likely to enable their bearers to survive and rear young which will in turn bear those characteristics. (See INCLUSIVE FITNESS.)

Flehmen German word describing a facial expression in which the lips are pulled back, head often lifted, teeth sometimes clapped rapidly together and nose wrinkled. Often associated with animals (especially males) sniffing scent marks or socially important odors (eg scent of estrous female). Possibly involved in transmission of odor to JACOBSON'S ORGAN.

Floe a sheet of floating ice.

Follicle a small sac, therefore (a) a mass of ovarian cells that produces an ovum, (b) an indentation in the skin from which hair grows.

Forestomach a specialized part of the stomach consisting of two compartments (presaccus and saccus).

Fossorial burrowing (of life-style or behavior).

Frugivore an animal eating mainly fruits.

Furbearer term applied to mammals whose pets have commercial value and form part of the fur harvest.

Gallery forest luxuriant forest lining the banks of watercourses.

Gamete a male or female reproductive cell (ovum or spermatozoon).

Gene the basic unit of heredity; a portion of DNA molecule coding for a given trait and passed, through replication at reproduction, from generation to generation. Genes are expressed as ADAPTATIONS and consequently are the most fundamental units (more so than individuals) on which NATURAL SELECTION acts.

Generalist an animal whose life-style does not involve highly specialized stratagems (cf SPECIALIST); for example, feeding on a variety of foods which may require different foraging techniques.

Genus (plural genera) a taxonomic division superior to species and subordinate to family (see TAXONOMY).

Gestation the period of development within the uterus; the process of **delayed implantation** can result in the period of pregnancy being longer than the period during which the embryo is actually developing. (See also IMPLANTATION.)

Glands (marking) specialized glandular areas of the skin, used in depositing SCENT MARKS.

Grizzled sprinkled or streaked with gray.

Harem group a social group consisting of a single adult male, at least two adult females and immature animals; the most common pattern of social organization among mammals.

Helper jargon for an individual, generally without young of its own, which contributes to the survival of the offspring of others by behaving parentally toward them (see ALLOPARENT).

Hemoglobin an iron-containing protein in the red corpuscles which plays a crucial role in oxygen exchange between blood and tissues in mammals.

Herbivore an animal eating mainly plants or parts of plants.

Hibernation a period of winter inactivity during which the normal physiological process is greatly reduced and thus during which the energy requirements of the animal are lowered.

Hierarchy (social or dominance) the existence of divisions within society, based on the outcome of interactions which show some individuals to be consistently dominant to others. Higher-ranking individuals thus have control of aspects (eg access to food or mates) of the life and behavior of low-ranking ones. Hierarchies may be branching, but simple linear ones are often called peck orders (after the behaviour of farmyard chickens).

Holarctic realm a region of the world including North America, Greenland, Europe, and Asia apart from the southwest, southeast and India.

Home range the area in which an animal normally lives (generally excluding rare excursions or migrations), irrespective of whether or not the area is defended from other animals (cf TERRITORY).

Hybrid the offspring of parents of different species.

Hyoid bones skeletal elements in the throat region, supporting the trachea, larynx and base of the tongue (derived in evolutionary history from the gill arches of ancestral fishes).

Implantation the process whereby the free-floating blastocyst (early embryo) becomes attached to the uterine wall in mammals. At

the point of implantation a complex network of blood vessels develops to link mother and embryo (the PLACENTA). In **delayed implantation** the blastocyst remains dormant in the uterus for periods varying, between species, from 12 days to 11 months. Delayed implantation may be obligatory or facultative and is known for some members of the Carnivora and Pinnipedia.

Inclusive fitness a measure of the animal's FITNESS which is based on the number of genes, rather than the number of its offspring, present in subsequent generations. This is a more complete measure of fitness, since it incorporates the effect of, for example, alloparenthood, wherein individuals may help to rear the offspring of their relatives (see KIN SELECTION: ALLOPARENT).

Induced ovulation see ESTRUS.

Infanticide the killing of infants. Infanticide has been recorded notably in species in which a bachelor male may take over a HAREM from its resident male(s).

Insectivore an animal eating mainly arthropods (insects, spiders).

Jacobson's organ a structure in a foramen (small opening) in the palate of many vertebrates which appears to be involved in olfactory communication. Molecules of scent may be sampled in these organs.

Juvenile no longer possessing the characteristics of an infant, but not yet fully adult.

Kin selection a facet of NATURAL SELECTION whereby an animal's fitness is affected by the survival of its relatives or kin. Kin selection may be the process whereby some ALLOPARENTAL behavior evolved; an individual behaving in a way which promotes the survival of its kin increases its own INCLUSIVE FITNESS, despite the *apparent* selflessness of its behavior.

Knuckle-walk to walk on all fours with the weight of the front part of the body carried on the knuckles; found only in gorillas and chimpanzees.

Lactation (verb: lactate) the secretion of milk from MAMMARY GLANDS.

Lanugo the birth-coat of mammals which is shed to be replaced by the adult coat.

Latrine a place where FECES are regularly left (often together with other SCENT MARKS); associated with olfactory communication.

Liana a climbing plant. In rain forests large numbers of often woody, twisted lianas hang down like ropes from the crowns of trees.

Lumbar a term locating anatomical features in the loin region (eg lumbar vertebrae are at the base of the spine).

Mammal a member of a CLASS of VERTEBRATE animals having MAMMARY GLANDS which produce milk with which they nurse their young (properly: Mammalia).

Mammary glands glands of female mammals that secrete milk.

Mangrove forest tropical forest developed on sheltered muddy shores of deltas and estuaries exposed to tide. Vegetation is almost entirely woody.

Mask colloquial term for the face of a mammal, especially a dog, fox or cat.

Matriline a related group of animals linked by descent through females alone.

Melanism darkness of color due to presence of the black pigment melanin.

Metabolic rate the rate at which the chemical processes of the body occur.

Migration movement, usually seasonal, from one region or climate to another for purposes of feeding or breeding.

Monogamy a mating system in which individuals have only one mate per breeding season.

Mutation a structural change in a gene which can thus give rise to a new heritable (genetic) characteristic.

Myoglobin a protein related to HEMOGLOBIN, found in the muscles of vertebrates; like hemoglobin, it is involved in the oxygen exchange processes of respiration.

Myopia short-sightedness.

Nasolacrimal duct a duct or canal between the nostrils and the eye.

Natal range the home range into which an individual was born (natal = of or from one's birth).

Natural selection the process whereby individuals with the most appropriate ADAPTATIONS are more successful than other individuals, and hence survive to produce more offspring. To the extent that the successful traits are heritable (genetic) they will therefore spread in the population.

Niche the role of a species within the community, defined in terms of all aspects of its life-style (eg food, competitors, predators, and other resource requirements).

Nocturnal active at nighttime.

Olfaction, olfactory the olfactory sense is the sense of smell, depending on receptors located in the epithelium (surface membrane) lining the nasal cavity.

Omnivore an animal eating a varied diet including both animal and plant tissue.

Opportunist (of feeding) flexible behavior exploiting circumstances to take a wide range of food items: characteristic of many species of Carnivora (see GENERALIST; SPECIALIST).

Order a taxonomic division subordinate to class and superior to family (see TAXONOMY).

Ovulation (verb: ovulate) the shedding of mature ova (eggs) from the ovaries where they are produced (see ESTRUS).

Pack ice large blocks of ice formed on the surface of the sea when an ice field has been broken up by wind and waves and drifted from its original position.

Pampas Argentinian steppe grasslands.

Papilla (plural: papillae) a small nipple-like projection.

Parturition the process of giving birth (hence *post partum*—after birth).

Pelvis a girdle of bones that supports the hindlimbs of vertebrates.

Perineal glands glandular tissue occurring between the anus and genitalia.

Pinna (plural: pinnae) the projecting cartilaginous portion the external ear.

Placenta, placental mammals a structure that connects the fetus and the mother's womb to ensure a supply of nutrients to the fetus and removal of its waste products. Only placental mammals have a well-developed placenta: marsupials have a rudimentary placenta or none, and monotremes lay eggs.

Plantigrade way of walking on the soles of the feet, including the heels (cf DIGITIGRADE).

Polyandrous see POLYGYNOUS.

Polygamous a mating system wherein an individual has more than one mate per breeding season.

Polygynous a mating system in which a male mates with several females during one breeding season (as opposed to polyandrous, where one female mates with several mates).

Population a more or less separate (discrete) group of animals of the same species within a given BIOTIC COMMUNITY.

Prairie North American steppe grassland between 30°N and 55°N.

Predator an animal which forages for live prey; hence "anti-predator behavior" describes the evasive actions of the prey.

Prehensile capable of grasping (eg of the tail).

Primary forest forest that has remained undisturbed for a long time and has reached a mature (climax) condition; primary rain forest may take centuries to become established.

Process (anatomical) an outgrowth or protuberance.

Promiscuous a mating system wherein an individual mates more or less indiscriminately.

Proximal near to the point of attachment or origin (eg the base of the tail).

Puberty the attainment of sexual maturity. In addition to maturation of the primary sex organs (ovaries, testes), primates may exhibit secondary sexual characteristics at puberty. Among higher primates it is usual to find a growth spurt at the time of puberty in both males and females.

Quadrumanous using both hands and feet for grasping.

Quadrupedal walking on all fours, as opposed to walking on two legs (bipedal) or moving suspended beneath branches in trees (suspensory movement).

Race a taxonomic division subordinate to SUBSPECIES but linking population with similar distinct characteristics.

Radiation see ADAPTIVE RADIATION.

Radio-tracking a technique used for monitoring an individual's movements remotely; it involves affixing a radio transmitter to the animal and thereafter receiving a signal through directional antennae, which enables the subject's position to be plotted. The transmitter is often attached to a collar, hence "radio-collar."

Rain forest tropical and subtropical forest with abundant and year-round rainfall. Typically species rich and diverse.

Receptive state of a female mammal ready to mate or in estrus.

Reduced (anatomical) of relatively small dimension (eg of certain bones, by comparison with those of an ancestor or related animals).

Reproductive rate the rate of production of offspring; the net productive rate may be defined as the average number of female offspring produced by each female during her entire lifetime.

Retractile (of claws) able to be withdrawn into protective sheaths.

Rhinarium a naked area of moist skin surrounding the nostrils in many mammals.

Rookery a colony of PINNIPEDS.

Ruminant a mammal with a specialized digestive system typified by the behavior of chewing the cud. The stomach is modified so that vegetation is stored, regurgitated for further maceration, then broken down by symbiotic bacteria. The process of rumination is an adaptation to digesting the cellulose walls of plant cells.

Savanna tropical grasslands of Africa, Central and South America and Australia. Typically on flat plains and plateaux with seasonal pattern of rainfall. Three categories—*savanna woodland, savanna parkland* and *savanna grassland*—represent a gradual transition from closed woodland to open grassland.

Scapula the shoulder-blade. Primates typically have a mobile scapula in association with their versatile movements in the trees.

Scent gland an organ secreting odorous material with communicative properties; (see SCENT MARK).

Scent mark a site where the secretions of SCENT GLANDS, or urine or FECES, are deposited and which has communicative significance. Often left regularly at traditional sites which are also visually conspicuous. Also the "chemical message" left by this means; and (verb) to leave such a deposit.

Scrub a vegetation dominated by shrubs— woody plants usually with more than one stem. Naturally occurs most often on the arid side of forest or grassland types, but is often artifically created by man as a result of forest destruction.

Seasonality (of births) the restriction of births to a particular time of the year.

Sebaceous gland secretory tissue producing oily substances, for example for lubricating and waterproofing hair, or specialized to produce odorous secretions.

Secondary forest (or growth) regenerating forest that has not yet reached the climax condition of PRIMARY FOREST.

Selective pressure a factor affecting the reproductive success of individuals (whose success will depend on their FITNESS, ie the extent to which they are adapted to thrive under that selective pressure).

Septum a partition separating two parts of an organism. The nasal septum consists of a fleshy part separating the nostrils and a vertical, bony plate dividing the nasal cavity.

Sinus a cavity in bone or tissue.

Solitary living on its own, as opposed to social or group-living in life-style.

Sonar sound used in connection with navigation (SOund NAvigation Ranging).

Specialist an animal whose life-style involves highly specialized stratagems: (eg feeding with one technique on a particular food).

Species a taxonomic division subordinate to genus and superior to subspecies. In general, a species is a group of animals similar in structure which are able to breed and produce viable offspring. See TAXONOMY.

Speciation the process by which new species arise in evolution. It is widely accepted that it occurs when a single-species population is divided by some geographical barrier.

Steppe open grassy plains of the central temperate zone of Eurasia or North America (prairies), characterized by low and sporadic rainfall and a wide annual temperature variation. In cold steppe, temperatures drop well below freezing point in winter, with rainfall concentrated in the summer or evenly distributed throughout year, while in hot steppe, winter temperatures are higher and rainfall concentrated in winter months.

Subadult no longer an infant or juvenile but not yet fully adult physically and/or socially.

Subfamily a division of a FAMILY.

Subfossil an incompletely fossilized specimen from a recent species.

Suborder a subdivision of an ORDER.

Subordinate see HIERARCHY.

Subspecies a recognizable subpopulation of a single SPECIES, typically with a distinct geographical distribution.

Taiga northernmost coniferous forest, with open boggy/rocky areas in between.

Tapetum lucidum a reflecting layer located behind the retina of the eye, commonly found in nocturnal mammals.

Taxonomy the science of classifying organisms. It is very convenient to group together animals which share common features and are thought to have common descent. Each individual is thus a member of a series of ever-broader categories (individual—species—genus—family— order—class—phylum) and each of these can be further divided where it is convenient (eg subspecies, superfamily or infraorder). The SPECIES is a convenient unit in that it links animals according to an obvious criterion, namely that they interbreed successfully. However, the unit on which NATURAL SELECTION operates is the individual; it is by the differential reproductive success of individuals bearing different characteristics that evolutionary change proceeds.

Terrestrial living on land.

Territory an area defended from intruders by an individual or group. Originally the term was used where ranges were exclusive and obviously defended at their borders. A more general definition of territoriality allows some overlap between neighbors by defining territoriality as a system of spacing wherein home ranges do not overlap randomly, that is, the location of one individual's or group's home range influences that of others.

Testosterone a male hormone synthesized in the testes and responsible for the expression of many male characteristics (contrast the female hormone estrogen produced in the ovaries).

Tooth-comb a dental modification in which the incisor teeth form a comb-like structure.

Tubercle a small rounded projection or nodule (eg of bone).

Tundra barren treeless lands of the far north of Eurasia and North America, on mountain tops and Arctic islands. Vegetation is dominated by low shrubs and herbaceous perennials, with mosses and lichens.

Underfur the thick soft undercoat fur lying beneath the longer and coarser hair (guard hairs).

Ungulate a member of the orders Artiodactyla (even-toed ungulates), Perissodactyla (odd-toed ungulates), Proboscidea (elephants), Hydracoidea (hyraxes) and Tubulidentata (aardvark), all of which have their feet modified as hooves of various types (hence the alternative name, hoofed mammals). Most are large and totally herbivorous (eg deer, cattle, gazelles, horses).

Vector an individual or species which transmits a disease.

Ventral on the lower or bottom side or surface: thus ventral abdominal glands occur on the underside of the abdomen.

Venule a small tributary conveying blood from the capillary bed to a vein (cf ARTERIOLE.)

Vertebrate an animal with a backbone: a division of the phylum Chordata which includes animals with notochords (as distinct from invertebrates).

Vestigial a characteristic with little or no contemporary use, but derived from one which was useful and well-developed in an ancestral form.

Vibrissae stiff, coarse hairs richly supplied with nerves, and with a sensory (tactile) function, found especially around the snout.

Xerophytic forest a forest found in areas with relatively low rainfall. Xerophytic plants are adapted to protect themselves against browsing (eg well-developed spines) and to limit water loss (eg small, leathery leaves, often with a waxy coating).

INDEX

A **bold number** indicates a major section of the main text, following a heading; a **_bold italic_** number indicates a fact box on a single species; a single number in (parentheses) indicates that the animal name or subjects are to be found in a boxed feature and a double number in (parentheses) indicates that the animal name or subjects are to be found in a spread special feature. _Italic_ numbers refer to illustrations.

Picture Acknowledgments

Key *t* top. *b* bottom. *c* center. *l* left. *r* right.

Abbreviations A Ardea. AN Agence Nature. BC Bruce Coleman Ltd. GF George Frame. J Jacana. FL Frank Lane Agency. OSF Oxford Scientific Films. FS Fiona Sunquist.

Cover BC, G. Ziesler. 1 BC, G. D. Plage. 2–3 BC, H. Jungius. 4–5 BC, M. P. Harris. 6–7 BC, Mary Grant. 8–9 BC, Leonard Lee Rue III. 11*t* GF. 11*b* T. N. Bailey. 14 M. Newdick. 15, 16 D. Macdonald. 17 T. P. O'Farrell. 20, 21 A. 22 BC. 23 GF. 24, 25 A. 26 GF. 27 B. Bertram. 28 World Wildlife Fund, A. Purcell. 29 Nature Photographers, M. Leach. 30 FS. 31 BC. 32*b* GF. 32*t* BC. 33 GF. 34*t* R. Caputo. 34*b*,

35 GF. 36 Natural Science Photos, G. Kinns. 37 A. 38, 39*tl*, 39*tr* GF. 39*b* BC. 40 AN. 41 BC, R. Williams. 42 Natural Science Photos. 43 Nature Photographers. 46–47 OSF. 51*t* G. 51*b* E. Zimen. 52–53 A. 53, 54*l* BC. 54*r*, 55 OSF. 56, 57*l* P. D. Moehlman. 57*r* A. 58–59 R. Caputo. 61 A. 62*t* BC. 62*b* FS. 63 OSF. 64, 64–65 A. 67 FS. 68 Anthro-Photo. 69*t* GF. 69*b* R. Caputo. 70 L. Malcolm. 71 GF. 73*t* A. J. T. Johnsingh. 73*b* E. R. C. Davidar. 74*l* BC. 74*r* J. Dietz. 75 BC, 76 A. Henley. 77*t* FS. 77*c* J. 77*b* BC. 78–79 J. W. Lentfer. 79, 80, 81, 82, 82–83 OSF. 84 F. Bruemmer. 85 J. W. Lentfer. 86 BC. 87*t* OSF. 87*b* A. 88 J. Mackinnon. 90 BC. 92 FL. 93*t* OSF. 93*b*, 94,

95 BC. 96, 96–97, 98 J. 99*t* BC. 99*c* J. 99*b* A. 102*t* BC, 102*b* A. 103*t* FL. 103*b*, 104, 105 BC. 107 FL. 108 Eric and David Hosking. 110*t*, *l* BC. 110*b* S. Buskirk. 111 J. 112 AN. 113 S. Carlsson. 114 BC. 116–117 J. 117 BC. 119 Survival Anglia. 120 N. Duplaix. 121 A. 123*t* FS. 123*b* J. 124 BC. 125 Survival Anglia. 126 A. E. Rasa. 127, 129 BC. 130 J. Mackinnon. 132 J. 133 BC. 134*t* AN. 134*b* J. 135 William Ervin, Natural Imagery. 138 D. Macdonald. 139 A. Henley. 140, 142–143 J. Rood. 144, 145 A. E. Rasa. 146 J. 148 G. Mills. 149*t* P. D. Moehlman. 149*b* GF. 150 FL. 151 P. Richardson.

Artwork

All artwork © Priscilla Barrett unless stated otherwise below.

Abbreviations JF John Fuller. OI Oxford Illustrators. SD Simon Driver.

12 JF. 13, 14 SD. 19*t* JF. 29, 30 VAP. 35, 45 OI. 49 JF. 52 John Brennan. 79 JF. 91*t* JF. 101*t* JF. 114*t* VAP. 127*t* JF. 147*t* JF. Maps and scale drawings SD.